CAMBRIDGE TRACTS IN MATHEMATICS

General Editors

B. BOLLOBAS, W. FULTON, A. KATOK, F. KIRWAN,
P. SARNAK

139   Typical Dynamics of Volume
       Preserving Homeomorphisms

Steve Alpern

London School of Economics

V. S. Prasad

University of Massachusetts, Lowell

Typical Dynamics of Volume Preserving Homeomorphisms

CAMBRIDGE
UNIVERSITY PRESS

PUBLISHED BY THE PRESS SYNDICATE OF THE UNIVERSITY OF CAMBRIDGE
The Pitt Building, Trumpington Street, Cambridge, United Kingdom

CAMBRIDGE UNIVERSITY PRESS
The Edinburgh Building, Cambridge CB2 2RU, UK   www.cup.cam.ac.uk
40 West 20th Street, New York, NY 10011-4211, USA   www.cup.org
10 Stamford Road, Oakleigh, Melbourne 3166, Australia
Ruiz de Alarcón 13, 28014, Madrid, Spain

First published 2000

Printed in the United Kingdom at the University Press, Cambridge

Typeface Computer Modern 10/13pt.   *System* LATEX 2_ε [DBD]

A catalogue record of this book is available from the British Library

ISBN 0 521 58287 3   hardback

Dedicated to the memory of John Oxtoby and Stan Ulam

Acknowledgements

We would like to thank all the people who helped us with this book, and earlier with the research on which it is based. Peter Lax and Jal Choksi are responsible for attracting us to this field and supervising our respective initial work in this area. John Oxtoby and Stan Ulam, to whom the book is dedicated, guided our subsequent investigations on which this book is based. Shizuo Kakutani encouraged us to extend our work to noncompact manifolds.

Much of the authors' collaboration on the book took place in Vermont and Boston, thanks to the hospitality of Judy and Saul Intraub and to Carla and Joe Wasserman. During the writing and proofreading of the book we benefitted from the help of several people, especially Ethan Akin, Ricardo Berlanga, and Fons Daalderop. The editing and publishing at Cambridge were expertly handled by Roger Astley and Sue Tuck. To these and others our sincere thanks for their contributions. On a personal note V.S.P. thanks his wife Mary for her support.

Contents

vii

Historical Preface

This monograph covers the authors' work over the past twenty five years on generalizing the classical results of John Oxtoby and Stan Ulam on the typical dynamical behavior of manifold homeomorphisms which preserve a fixed measure. In the main text of the book we will take a logical rather than historical perspective, designed to give the reader a concise and unified treatment of results we obtained in a series of articles that were written before the overall structure of the theory was clear. However, since the true significance of this field of study can be understood only from a historical perspective, we devote this preface to a discussion of the problem considered by Oxtoby and Ulam when they were Junior Fellows at Harvard in the 1930s, and of their accomplishment in its solution. We shall use their own words where possible.

The origins of Ergodic Theory lie in the study of physical systems which evolve in time as solutions to certain differential equations. Such systems can be initially described by parameters giving the states of the system as points in Euclidean n-space. Taking conservation laws into account, the phase space may be decomposed into lower dimensional manifolds. Regularities in the differential equations obeyed by the system are reflected in the differentiability or the continuity of the flow that describes the evolution of the system over time. Furthermore, Liouville's Theorem ensures that for Hamiltonian systems this flow has an invariant measure. Thus one is led in a natural way from the underlying physics to the study of measure preserving manifold homeomorphisms or diffeomorphisms. As the latter have received much attention we will confine our attention here to the case of homeomorphisms.

An important historical assumption that was often made in the study of such systems was the so called 'ergodic hypothesis' of statistical mechanics, as described by Oxtoby and Ulam in their 1941 paper [88, p. 874]:

In the classical theory the assumption was made that the average time spent in any region of phase space is proportional to the volume of the region in terms of the invariant measure, more generally, that time-averages may be replaced by space-averages. To justify this interchange, a number of hypotheses were proposed, variously known as ergodic or quasi-ergodic hypotheses. ... A rigorous discussion of the precise conditions under which the interchange was admissible was only made possible in 1931 by the ergodic theorem of Birkhoff. This established the *existence* of the time-averages in question ... and showed that ... the interchange is permissible if and only if the flow in phase space is *metrically transitive* [the older term for *ergodic*]. A transformation or a flow is metrically transitive [ergodic] if there do not exist two disjoint invariant sets both having positive measure. Thus the effect of the ergodic theorem was to replace the ergodic hypothesis by the hypothesis of metrical transitivity [ergodicity].

An important question in the 1930s was consequently the determination of which known transformations were ergodic, and more generally, which manifolds could support an ergodic homeomorphism. Aside from the pure existence question, both Birkhoff and Hopf had conjectured that ergodicity was the *general case* for transformations, in some unspecified sense. A natural setting at that time in which to make their conjecture precise was Baire's notion of category. In this topological context, ergodic homeomorphisms represent the general case if the nonergodic ones constitute a set of *first category* (that is, the union of countably many nowhere dense sets).

When Oxtoby and Ulam were Junior Fellows at Harvard in the late 1930s, the main problem they worked on was the determination of those (connected) compact manifolds for which ergodicity was the general case for measure preserving homeomorphisms. Their main finding was that ergodicity is the general case for *all* compact manifolds, or as they put it, 'the hypothesis of metrical transitivity in dynamics involves no *topological* contradiction'. John Oxtoby told us that during this period G. D. Birkhoff was their main source of problems (in particular this one) and Marshall Stone was the main source of techniques regarding their solution.

Ulam describes his work with Oxtoby on this problem in his autobiography *Adventures of a Mathematician* [103], in the chapter Harvard Years, 1936–1939:

In order to complete the foundation of the ideas of statistical mechanics connected with the ergodic theorem, it was necessary to prove the existence, and what is more, the prevalence of ergodic transformations. G. D. Birkhoff himself had worked on special cases in dynamical problems, but there were no general results. We wanted to show that on every manifold (a space representing the possible states of a dynamical system) – the kind used in statistical mechanics – such ergodic behavior is the rule. ... We discussed various approaches to a possible construction of these transformations. ... We kept G. D. Birkhoff informed of the status of our attacks on the problem. ... He would check what I told him with Oxtoby, a more cautious person. It took us more than two years to break through and to finish a long paper [88] which appeared in *The Annals of Mathematics* in 1941, and which I consider one of the more important results that I had a part in.

The result of Oxtoby and Ulam that ergodicity is generic for measure preserving homeomorphisms of compact manifolds has been generalized in two ways. The first direction in which their result extends is that the property of ergodicity has been generalized to more specialized measure theoretic behavior. This was first done by Katok and Stepin [76], who in 1970 proved that weak mixing homeomorphisms are also generic. To put Katok and Stepin's result in a historical context, we note that subsequent to Oxtoby and Ulam's 1941 paper, Paul Halmos published two papers: the first [69] in 1944 showed that ergodicity is generic in the weak topology in the space of all measure preserving bijections (called automorphisms) of a measure space; in a second paper that year [70], Halmos proved that weak mixing is also a generic property for measure preserving bijections. In describing the relation between his theorem [69, Theorem 6] on ergodicity being generic for measure preserving bijections and Oxtoby and Ulam's theorem on generic ergodicity for measure preserving homeomorphisms, Halmos notes [69, p. 2, footnote 1]:

The first theorem of this type is due to J. C. Oxtoby and S. M. Ulam ... Their topology is however, very different from mine and depends on the topological and metric (as opposed to purely measure theoretic) structure of the underlying space.

Further on in his paper, Halmos states [69, p. 12]:

... there is, however, no implication between [Halmos's] Theorem 6 and the corresponding result of Oxtoby and Ulam: they define a stronger topology and I consider a wider class of transformations.

Halmos's statement notwithstanding, the first author (S. Alpern) showed that in fact any measure theoretic property which is generic for abstract measure preserving automorphisms is also generic for measure preserving homeomorphisms of compact manifolds. Thus Alpern's result related the two 1944 papers of Halmos in the former context (proofs that ergodicity and then weak mixing were generic) to the work of Oxtoby–Ulam and Katok–Stepin. This generalization of the Oxtoby–Ulam Theorem to all typical measure theoretic properties is covered in the first half of the book (Parts I and II), which is devoted to compact manifolds. In fact most of the theory is developed in Part I in the special context of volume preserving homeomorphisms of the unit n-cube. Part II shows how these results may be generalized to homeomorphisms of a compact manifold which preserve a certain finite measure. Some of the more elementary aspects of this work can be very simply developed using the ideas of Lax [80] on discrete approximation of measure preserving homeomorphisms, including some applications to fixed point theory. However, the main logical development is independent of these combinatorial notions and uses instead the idea of viewing the space of measure preserving *homeomorphisms* of a manifold as being embedded in the larger space consisting of all *bijections* of the manifold which preserve that measure. Properties of this embedding are established through a Lusin Theorem for measure preserving homeomorphisms.

The second direction of generalization of the result of Oxtoby and Ulam, covered in Part III, is the removal of the compactness assumption on the underlying manifold, and the concomitant consideration of infinite preserved measures. Although Besicovitch had established the existence of a transitive homeomorphism of the plane in 1937, the corresponding result for ergodicity was not established until 1979, when Prasad [96] showed that in fact ergodicity is generic for volume preserving homeomorphisms of R^n. However, it soon became clear that unlike the compact case, in which all manifolds supported generic ergodicity, not all noncompact manifolds had this property. The search for the relevant manifold property which determined the supported dynamical behavior then centered on the so called *ends* of the manifold, roughly speaking, the distinct ways of going to infinity. The purely measure theoretic underpinning for the infinite measure work was established by Choksi and Kakutani [50], who showed in 1979 that ergodicity is a typical property for measure preserving bijections of an *infinite* Lebesgue space.

For noncompact manifolds, the space of measure preserving homeomorphisms divides into components according to the induced homeomorphism of the set of ends. We find, for example, that if the induced end homeomorphism is transitive then ergodicity is generic within such a component. Furthermore, if the induced end homeomorphism is topologically weak mixing, then any property generic for measure preserving transformations of an infinite Lebesgue space is generic within the component. A fuller description of the authors' work on noncompact manifolds is contained in the Introduction to Part III.

This book covers only those aspects of the field of measure preserving homeomorphisms of a manifold that involve *typical* properties of such transformations. So for example we do not discuss the important result of Lind and Thouvenot [83] on ergodic theoretic behavior represented by some measure preserving torus homeomorphisms, because the behavior they demonstrate is not typical.

Our aim is to give a streamlined approach to our work in this area, from a perspective only recently reached and not fully appreciated in our articles on the subject. As this is a work centered on the interaction of measure and topology, we have given full proofs of all results that combine these two fields (the core of the theory) while leaving out some proofs of results that fall fully within measure theory or manifold topology.

Most of the work described in the first two parts of the book was carried out under the guidance and encouragement of John Oxtoby. The early work of Alpern in this area also benefited from discussions with Stan Ulam. Aside from these two founders of the field of measure preserving homeomorphisms, the four mathematicians whose ideas most influenced this work are Jal Choksi, Robert D. Edwards, Shizuo Kakutani, and Peter Lax.

General Outline

The book as a whole gives a unified presentation of the authors' work on establishing conditions under which an ergodic theoretic dynamical property is typical in the space $\mathcal{M}[X, \mu]$ consisting of all homeomorphisms of a sigma compact manifold X which preserve a fixed nonatomic Borel measure μ. The first half of the book, comprising Parts I and II, covers the first author's work on compact manifolds (for which μ is necessarily finite). For clarity of exposition the material in the first eight chapters (Part I) is presented for the special case where the compact manifold X is simply the unit n-dimensional cube I^n and the measure μ is n-dimensional Lebesgue measure (volume). In Part II, comprising Chapters 9 and 10, we show how the results obtained for the cube hold as well for arbitrary compact manifolds. The second half of the book, Part III (Chapters 11–17), describes the work of both authors in extending the earlier work to the case where the manifold X is not compact (and μ may be infinite). In some cases the earlier work for the compact case cannot be extended, and we establish such negative results as well. In this half of the book the results depend in a significant way on the structure of the 'ends' of the manifold X, which are roughly the ways of going to infinity on the manifold. In particular, the ergodic theoretic properties of a μ-preserving homeomorphism h of a noncompact manifold X will depend on its induced action on the ends of X and on the net measure that it flows into each end. Following Part III, there are two appendices. Appendix 1 is mainly concerned with presenting a purely measure theoretic result of the first author, which we call the Multiple Tower Rokhlin Theorem, as it generalizes a similar result due to Rokhlin and Halmos for a single tower. Corollaries of this theorem, as well as that of an infinite measure version due to the authors and J. Choksi, are used extensively in the main part of the book. Appendix 2 is the only chapter of the book which is not based on the work of the authors. It presents theorems, due to von Neumann and Oxtoby–Ulam (for compact manifolds), to Oxtoby–Prasad (for the Hilbert cube), and to Berlanga and Epstein (for sigma compact manifolds), which give necessary and sufficient conditions for two measures μ and ν on a manifold to be 'homeomorphic'. This means that for some self-homeomorphism of the manifold we have $\mu(A) = \nu(h(A))$ for all Borel sets A.

Recalling the first sentence of this outline, we now say what we mean by an 'ergodic theoretic dynamical property' and by 'typical'. There are of course many types of properties that a measure preserving homeomorphism might possess. For example if the manifold is the cube, it must have a fixed point. However, this property 'lives' on a set of measure zero, and we are concerned mainly with properties that 'live' on a set of full measure, such as ergodicity. More precisely, we are concerned with properties that can be defined in the larger space $\mathcal{G}[X,\mu]$ consisting of all μ-preserving bijections of the manifold X viewed simply as a measure space, where the manifold structure is irrelevant. Examples of such measure theoretic properties are ergodicity, weak mixing, and zero entropy. In order to say what we mean by a 'typical' property, we must endow the space $\mathcal{M}[X,\mu]$ with a topology, which we take to be the uniform topology when X is compact, and more generally the topology of uniform convergence on compact sets when it is not. Then we say a property is typical in $\mathcal{M}[X,\mu]$ if the homeomorphisms possessing it contain a dense G_δ subset. In a similar fashion we will say that a measure theoretic property is typical in the space $\mathcal{G}[X,\mu]$ if it contains a dense G_δ subset of that space, with respect to a commonly used topology called the weak topology.

The main aim of the first half of the book, carried out in Parts I and II, is the derivation of the first author's result that any measure theoretic property (such as ergodicity or weak mixing) which is typical in the measure theoretic context (that is, in $\mathcal{G}[X,\mu]$) is also typical for measure preserving homeomorphisms (that is, in $\mathcal{M}[X,\mu]$ for compact manifolds X). The aims of the second half are more varied. We develop the results of both authors in establishing positive and negative results regarding the typicality of certain properties on various manifolds. Unlike the fairly universal result stated above for compact manifolds (universal in that all properties and all manifolds are treated similarly), we find that our results for noncompact manifolds depend both on the property and on the manifold. For Euclidean space R^n, the second author showed that ergodicity is typical for volume preserving homeomorphisms. This is the first result presented in Part III. We then present examples of manifolds where ergodicity is not typical. After an extensive treatment of the interaction of ends and measures, we obtain a synthesis of the positive and negative results regarding ergodicity: A homeomorphism h in $\mathcal{M}[X,\mu]$ is the limit of ergodic homeomorphisms if and only if it does not compress any set of ends of X (into a proper subset of itself) and it does not induce a positive flow of measure into any set of ends. As the identity homeomorphism on X has these properties, it follows that any manifold supports an ergodic homeomorphism. We then consider more general properties, and show that any property typical in $\mathcal{G}[X,\mu]$ is typical in a certain nonempty closed subspace of $\mathcal{M}[X,\mu]$, and is consequently possessed by some μ-preserving homeomorphism of X. In particular there are weak mixing homeomorphisms of any sigma compact manifold (X,μ).

Despite our earlier disclaimer regarding properties that live on sets of measure zero, the book does include results for such a property, namely *maximal chaos*. This is a topological property introduced by the authors which entails topological transitivity, dense periodic points, and a maximal form of sensitive dependence on initial conditions. As such, it is a strictly stronger property than Devaney's version of chaos. In Chapter 4 we establish that homeomorphisms with maximal chaos are dense in $\mathcal{M}[X,\mu]$ when X is compact, and in Chapter 17 we establish that for arbitrary sigma compact manifolds such homeomorphisms are dense in a nonempty subset of $\mathcal{M}[X,\mu]$. In particular, any sigma compact manifold supports a maximally chaotic homeomorphism. In addition to the topological property of chaos, we also apply our techniques to the fixed point property. In Chapter 5 we look at the relationship between this property and area preservation, for various 2-dimensional manifolds. We apply an approximation technique due to Peter Lax to give simple proofs of both the Poincaré–Birkhoff Theorem and the Conley–Zehnder–Franks Theorem, results which assert the existence of fixed points for area preserving homeomorphisms of the annulus and torus, respectively, under some additional hypotheses. The same technique (involving the Marriage Theorem) is also used in Chapter 7 to give a new proof that ergodicity is typical for volume preserving homeomorphisms of the cube.

The material above is meant to give the reader a very informal idea of the main results covered in this book. For a slightly more detailed presentation of the main results, the reader is referred to Section 1.3

(for results on compact manifolds) and Section 11.3 (for noncompact manifolds).

We wish to assure readers who come to this book with little or no familiarity with the fields of ergodic theory or measure theory that no prior knowledge of these fields is required. All the ergodic theoretic notions that we will use will be explained and defined when they are needed.

Part I
Volume Preserving Homeomorphisms of the Cube

1

Introduction to Parts I and II
(Compact Manifolds)

1.1 Dynamics on Compact Manifolds

Two of the principal analytic structures that may be put on a set X are measure and topology. We are interested in transformations of X which preserve *both* of these structures: measure preserving homeomorphisms. In the first half of the book, Parts I and II, the topological space X will be a compact manifold, possibly with boundary. (In fact Part I specializes to the case where X is simply the unit cube I^n in some dimension $n \geq 2$.) The measure, denoted μ, will be a nonatomic Borel probability measure which assigns the manifold boundary measure zero and is positive on all nonempty open sets (a property we call *locally positive*). (In Part I, μ is simply the volume measure on the cube.) The first two parts of the book are concerned with determining typical properties of μ-preserving homeomorphisms of the (arbitrary) compact manifold X. We denote the set of all such homeomorphisms by $\mathcal{M}[X,\mu]$, which we endow with the uniform topology, with respect to which it is complete. We call a property *typical*, or *generic*, if it is possessed by a dense G_δ (or larger) subset of transformations. The purpose of this introductory chapter is to give a nontechnical presentation of the main results, and the definitions they involve, for measure preserving homeomorphisms of compact manifolds. Both the definitions and theorems mentioned in this chapter will be presented more rigorously in later chapters.

1.2 Automorphisms of a Measure Space

Given X and μ, we will often consider more general transformations called *automorphisms*, which are bimeasurable bijections of X which preserve the measure μ. In particular, automorphisms do not need to

3

be continuous. Since the topological structure of X is ignored the remaining measure space (X, μ) is measure theoretically the same as the unit interval with Lebesgue measure. Such a measure space is called a *finite Lebesgue space* (see [71]). Consequently the space $\mathcal{G} = \mathcal{G}[X, \mu]$ consisting of all automorphisms of (X, μ) is essentially the same as the space of all Lebesgue measure preserving bijections of the unit interval. We endow the space $\mathcal{G}[X, \mu]$ with the *weak topology*, which is determined by defining the sequential convergence of a sequence of automorphisms g_i to a limit automorphism g if $\mu\left(g_i(A) \bigtriangleup g(A)\right) \to 0$ for all measurable sets A. Here the symbol $\bigtriangleup$ denotes the *symmetric difference* between sets, defined by $A \bigtriangleup B = \left(A \cap \tilde{B}\right) \cup \left(\tilde{A} \cap B\right) = (A - B) \cup (B - A)$. The space $\mathcal{G}[X, \mu]$ is complete with respect to the weak topology.

1.3 Main Results for Compact Manifolds

Historically, the question of typical properties has been studied quite separately for the two settings $(\mathcal{G}[X, \mu],$ weak topology) and $(\mathcal{M}[X, \mu],$ uniform topology) with different techniques being applied. In each case, the first property shown to be typical was *ergodicity*. (An automorphism of a measure space is called ergodic if every invariant set either has measure zero or its complement has measure zero.) Ergodicity was proved to be typical for $\mathcal{G}[X, \mu]$, that is for automorphisms of any finite Lebesgue space, by Halmos in 1944 [69]. This followed the slightly earlier (1941) and more difficult proof of Oxtoby and Ulam [88] that ergodicity is typical in $\mathcal{M}[X, \mu]$. In a second 1944 paper, Halmos further proved that *weak mixing* automorphisms are typical in $\mathcal{G}[X, \mu]$ (an automorphism f of (X, μ) is weak mixing if $f \times f$ is ergodic on $(X \times X, \mu \times \mu)$). However, it was not until 1970 that Katok and Stepin [76] proved the corresponding result for $\mathcal{M}[X, \mu]$. Other properties have also been shown to be typical in both spaces, first in $\mathcal{G}[X, \mu]$ and later in $\mathcal{M}[X, \mu]$. In the case of homeomorphisms these results are also existence results for the specified measure theoretic behavior on arbitrary compact manifolds, since it is not known how to construct examples. However, it is easy to construct automorphisms with the required behavior. The main purpose of this part of the book is to unify these two theories in the following Theorem C obtained by the first author in 1978 [11]. In the form given, it is Corollary 10.4, which follows from a symmetric version giving simultaneous typicality in both contexts (Theorem 10.3). By a 'measure theoretic property', we mean a set $\mathcal{V}$ of automorphisms which is invariant under conjugation by any automorphism (i.e., $\mathcal{V} \subset \mathcal{G}[X, \mu]$ such that

$g^{-1}\mathcal{V}g = \mathcal{V}$ for all $g \in \mathcal{G}[X, \mu]$). See also Theorem 8.2 for a version of the theorem for the cube.

Theorem C *If a measure theoretic property is typical for length preserving automorphisms of the unit interval then it is also typical for homeomorphisms of a compact manifold which preserve a given finite nonatomic locally positive measure.*

The main idea of this part of the book, used to obtain the unification mentioned above in Theorem C, is to view the space $\mathcal{M}[X, \mu]$ as a subset of $\mathcal{G}[X, \mu]$. Thus even when the questions are entirely about homeomorphisms in $\mathcal{M}[X, \mu]$, we may employ approximations which go outside that space into $\mathcal{G}[X, \mu]$ and hence do not have to be continuous.

In order to obtain Theorem C, we need two results on the embedding of $\mathcal{M}[X, \mu]$ in $\mathcal{G}[X, \mu]$. The first, Theorem A (Theorem 8.4), lets us uniformly approximate any homeomorphism in $\mathcal{M}[X, \mu]$ by an automorphism with a desired measure theoretic property (e.g., weak mixing). An automorphism is called antiperiodic if its set of periodic points has zero measure.

Theorem A (Conjugacy Approximation) *Any homeomorphism in $\mathcal{M}[X, \mu]$ may be uniformly approximated by an automorphism of the underlying measure space which is conjugate to any given antiperiodic automorphism.*

For example, if the given automorphism is taken to be ergodic, this says that any μ-preserving homeomorphism may be uniformly approximated by an ergodic automorphism. However, since the approximating automorphism is not necessarily continuous (may lie outside $\mathcal{M}[X, \mu]$), we need an additional mechanism to eventually go back into the space $\mathcal{M}[X, \mu]$ of homeomorphisms. The relevant mechanism is a type of Lusin Theorem which says that

Theorem B (Lusin Theorem for Measure Preserving Homeomorphisms) *The space $\mathcal{M}[X, \mu]$ is dense in the space $\mathcal{G}[X, \mu]$, in the weak topology.*

Actually a stronger version of this result, Theorem 6.2, is needed. These two results (Theorem A (8.4) and Theorem B (6.2)) on the embedding of $\mathcal{M}[X, \mu]$ in $\mathcal{G}[X, \mu]$ are exactly what is needed to obtain the synthesis of Theorem C (Corollary 10.4) mentioned above regarding the identity of typical measure theoretic properties in the two spaces. These three results, Theorems A, B, C, form the core of this half of the book, on compact manifolds. In addition, we make extensive use of the 'Homeomorphic Measures Theorem' of von Neumann, and Oxtoby and Ulam, which enables us to restrict ourselves to the simple case of the unit cube with Lebesgue measure for the first eight chapters (which we call Part I), and then to simply extend the theory in Chapters 9 and 10 (which we call Part II) to any finite nonatomic locally positive measure on any compact manifold. Thus the core of this half of the book is contained in Chapters 2 (definitions), 6 (Theorem B), 8 (Theorems A, C), and 9, 10 (covering the applications of the Homeomorphic Measures Theorem).

All of these theorems establish typical ergodic theoretic behavior for volume preserving homeomorphisms. Some of the techniques can be used to establish some typical topological dynamical properties for volume preserving homeomorphisms such as transitivity or chaos. Theorem 4.8 shows that every volume preserving homeomorphism of the n-cube ($n \geq 2$) can be uniformly approximated by one which is maximally chaotic (the latter notion is stronger than the usual notion of chaos in the sense of Devaney – see Chapter 4). This result can be combined with a result of Daalderop and Fokkink [55] to prove

Theorem D   *Maximal chaos is typical for volume preserving homeomorphisms of the cube.*

This is a purely topological result which has no counterpart in $\mathcal{G}[I^n, \lambda]$, the space of volume preserving automorphisms.

In addition to the above core results of this half of the book, we present a number of ancillary results based on Peter Lax's idea of approximating volume preserving homeomorphisms of the cube by permutations of the cells of some dyadic decomposition. This is a very powerful and intuitive technique which often lead to the initial proofs of new results. Indeed, the first (slightly weaker) versions of Theorems A, B, C were based on this combinatorial idea. For this reason we have included a chapter on these combinatorial techniques, as well as chapters on some applications: existence of a transitive homeomorphism of the cube and of R^n, a proof of Poincaré's Last Geometric Theorem, and the typicality of ergodicity and chaos for volume preserving homeomorphisms of the cube. The results of these Chapters (3, 4, 5, 7) will not be used elsewhere, so these chapters may be considered optional. However, they will certainly increase the reader's intuitive grasp of the ideas in this book.

2

Measure Preserving Homeomorphisms

2.1 The Spaces $\mathcal{M}, \mathcal{H}, \mathcal{G}$

This book is primarily concerned with typical measure theoretic properties (such as ergodicity or weak mixing) of members of the space $\mathcal{M}[X, \mu]$ consisting of all self-homeomorphisms of a manifold X which preserve a given Borel measure μ. To a much lesser extent, we will also consider topological properties, such as transitivity or the existence of fixed points. We will only consider manifolds of dimension at least 2. The transformations we study preserve both the measure theoretic and topological structure of the underlying space. That is, they belong to both the space of self-homeomorphisms of the manifold (denoted $\mathcal{H}[X]$) and to the space of *automorphisms* of the underlying Borel measure space (X, μ) (denoted $\mathcal{G}[X, \mu]$). An automorphism $g \in \mathcal{G}[X, \mu]$ is a bijection $g : X \to X$ with both g and g^{-1} measurable and $\mu(A) = \mu(g(A)) = \mu(g^{-1}(A))$ for all measurable sets A. Automorphisms which differ on a set of measure zero will be identified. The measure theoretic properties that we are interested in, such as ergodicity and weak mixing, do not rely on the topology of the underlying space – rather they depend only on the measure theoretic structure of the space, and for the manifolds we consider these are all the same: namely the manifolds that we consider are all measure theoretically the same as the standard Lebesgue space (I, λ), the unit interval with the sigma algebra of Lebesgue measurable sets and Lebesgue measure λ (length measure). Such measure spaces (X, μ) are called Lebesgue spaces and are distinguished only by their total measure $\mu(X)$.

In Parts I and II we consider compact manifolds with probability measures and indeed for Part I (Chapters 1–8) we consider only Lebesgue measure λ (n-dimensional volume measure) on the n-cube I^n. We denote

7

by $\mathcal{M}[I^n, \lambda]$ the space of volume preserving homeomorphisms of the unit cube I^n. In Part II, Chapters 9 and 10 we will show that all the results obtained for this special case can be easily extended to compact manifolds with finite nonatomic measures which are positive on open sets.

In the compact case we will endow the spaces $\mathcal{H}[X]$, $\mathcal{M}[X,\mu]$, and $\mathcal{G}[X,\mu]$ with the *uniform topology* defined by the metric $\|f - g\| \equiv$ $\operatorname{ess\,sup}_{x \in X} d(f(x), g(x))$, where d is a metric on the manifold X, usually the Euclidean or maximum metrics on the cube or torus, and denoted by $|x - y|$. We will also denote $\|f\| \equiv \operatorname{ess\,sup}_{x \in X} d(f(x), x)$ as the *norm* of f, observing that $\|fg^{-1}\| = \|f - g\|$ in our notation. Of course for the spaces $\mathcal{H}[X]$, $\mathcal{M}[X,\mu]$, the essential supremum reduces to the maximum. We will use the notation $\mathcal{H}[X, Y]$ ($\mathcal{M}[X, Y, \mu]$) to denote the subspace of $\mathcal{H}[X]$ (respectively $\mathcal{M}[X,\mu]$) consisting of homeomorphisms equal to the identity on the subset Y.

The spaces $\mathcal{H}[X]$ and its closed subset $\mathcal{M}[X,\mu]$ are not complete under the uniform topology metric given above. However, they are *topologically complete*, since that metric is equivalent to the complete metric defined by $u(f, g) = \|f - g\| + \|f^{-1} - g^{-1}\|$ (see [91]). We call this the *uniform metric*. This will justify our repeated application of the Baire Category Theorem (see [91] for discussion and proof):

Theorem 2.1 *In a complete metric space the countable intersection of dense open sets is dense.*

A set which is the countable intersection of open sets is called a G_δ set. We shall call a property *typical*, or *generic*, if the set of points with this property contains a dense G_δ set. Typical properties represent sets which are large in a topological sense, and in particular, nonempty. For this reason many of the results to be presented here can be considered existence proofs. For example, the main classical result of Oxtoby and Ulam says that ergodicity is typical among measure preserving homeomorphisms of a compact manifold. We note that a set $V \subset X$ is *nowhere dense* if for every nonempty open set U there is a nonempty open set in $U - V$ (i.e., every open set U has an open subset, 'a hole', missing V). It is easy to see that V is nowhere dense if and only if the interior of the closure of V is empty. Thus a nowhere dense set V is a 'topologically small' set (V is like a piece of Swiss cheese where the 'holes' missing V are dense in every open set). Furthermore, in a complete metric space the countable union of closed nowhere dense sets is small (since by the

Category Theorem, the complement would be a dense G_δ set). Any set which is the countable union of closed nowhere dense sets in a complete metric space is said to be a *set of first (Baire) category* (or *Baire category I*) – the complement of a set of first category is called a *residual set*. For a delightful investigation of the analogies between notions of topological smallness and measure theoretic smallness (zero measure) see J. C. Oxtoby's book *Measure and Category* [91].

At this point in the exposition, the reader would probably like to see some examples of measure preserving homeomorphisms. There is always of course the identity map. On manifolds with an additive structure which leaves the measure invariant (e.g., Euclidean space or the torus), translations of the form $x \mapsto x + c$ give simple examples. Unfortunately these will be of no use to us on general manifolds, or on the important special case of the cube, because they cannot be localized. On Euclidean space, rotations form another important example. These will in fact be useful in general because they can be localized. For example, given a planar disk of radius r centered about a point p, and a continuous function $\alpha : [0, r] \to [0, \infty)$, $\alpha(r) = 0$, we may consider the transformation which rotates the circle of radius t by an angle $\alpha(t)$, for $t \leq r$. We call this a *variable rotation*. This is clearly an area preserving homeomorphism, and we shall find that most of our constructions are ultimately limits of compositions of such local variable rotations. (An exception to this is the construction in Chapter 6.)

2.2 Extending a Finite Map

A simple question one may ask about the space $\mathcal{M}[I^n, \lambda]$ of volume preserving homeomorphisms of the cube, is whether it acts transitively on the interior. That is, given any pair of interior points x, y, can one always find a transformation h in $\mathcal{M}[I^n, \lambda]$ with $h(x) = y$? Actually, the space $\mathcal{M}[I^n, \lambda]$ possesses the stronger *finite extension property*: Any embedding $\check{h} : F \to I^n$ of a finite set $F \subset \text{Int } I^n$, the interior of I^n, can be extended to a homeomorphism h in $\mathcal{M}[I^n, \lambda]$ with the norm $\|h\|$ as close to that of $\check{h}$ as desired. It is this property that allows us to combine combinatorial constructions based on finite sets with continuous approximations of various sorts. The actual construction of the extension h outlined in the lemmas below is slightly more explicit than in the original proof of Oxtoby and Ulam, to allow some additional applications not given in their original paper (in particular to the Lusin theory given in Chapter 6). It uses the variable rotations discussed in the previous section.

Lemma 2.2 *Given any two points p, q in R^n $(n \geq 2)$, and any positive number δ, let $B = B(p, q; \delta)$ denote the closed Euclidean ball centered at the midpoint of p and q, and with radius $|p-q|/2+\delta$, where $|x-y|$ denotes Euclidean distance. Then there is a volume preserving homeomorphism h of R^n which equals the identity off B, satisfies $h(p) = q$, and maps some neighborhood of p rigidly onto a neighborhood of q.*

Proof First consider the case $n = 2$. Define h by rotating the disk $B(p, q; \delta/2)$ by an angle π and rotating the circle given by the boundary of $B(p, q; \delta/2+t)$ by the angle $\pi - 2\pi t/\delta$, for $0 \leq t \leq \delta/2$. For $n > 2$, let $D = D(p, q; \delta)$ be the intersection of $B(p, q; \delta)$ with any 2-dimensional plane through p and q. Define h on D as in the 2-dimensional case and extend it to $B(p, q; \delta)$ by requiring it to be a rigid motion of every sphere concentric to $B(p, q; \delta/2)$. Finally, extend h to the rest of R^n by setting it equal to the identity off B. $\qquad\square$

Lemma 2.3 *Let $U \subset I^n$ be an open neighborhood of an arc L from p to q. Then there is a volume preserving homeomorphism h of I^n with $h(p) = q$, which maps some neighborhood of p onto a neighborhood of q by simple translation, and equals the identity off U.*

Proof Choose $\delta > 0$ and points $p = p_0, p_1, \ldots, p_k = q$ in L sufficiently close so that $B(p_i, p_{i+1}; \delta) \subset U$, for $i = 0, \ldots, k-1$. For $i = 0, \ldots, k-1$, let h_i be the homeomorphism given by the previous lemma for the points p_i and p_{i+1}. Then the composition $h = h_k \circ h_{k-1} \circ \cdots \circ h_1 \circ h_0$ will be the required homeomorphism if the homeomorphism h_k is an appropriate rigid motion of $B(q, q; \delta)$ which equals the identity off U. $\qquad\square$

Theorem 2.4 *Let $\{p_i\}_{i=1}^N$ and $\{q_i\}_{i=1}^N$ be two sets of N distinct interior points of I^n, with $|p_i - q_i| < \epsilon$. Then there is a volume preserving homeomorphism h, with $\|h\| < \epsilon$ and equal to the identity on the boundary of I^n, which for each i maps some neighborhood of p_i by simple translation onto a neighborhood of q_i, and p_i into q_i. The homeomorphism h can be made to equal the identity on a given finite set disjoint from the p's and q's.*

This result also holds for any manifold possessing a metric with the property that any two points at a distance less than δ can be joined *by an arc of length less than δ (the underlying metric will be denoted by $|x - y|$). Note that the maximum metric (on I^n or the torus T^n) has this property. In all cases the required homeomorphism h can be constructed as the composition of homeomorphisms with support $(\{x : h(x) \neq x\})$ in balls.*

Proof Select N arcs $L_i : [0, 1] \to \text{Int } I^n$, satisfying

$$d(L_i(t), L_i(t')) < \epsilon |t - t'|, \ L_i(0) = p_i, \ L_i(1) = q_i.$$

We first prove the result under the assumption that the N sets $L_i [0, 1]$ are disjoint, and then use this special case to prove the general result.

Assuming disjoint arcs L_i, we may choose a $\delta > 0$ sufficiently small so that the sets U_i of points within distance δ of the arc L_i are disjoint open subsets of the interior of I^n, with diameter less than ϵ. Applying the previous lemma for each i, we obtain volume preserving homeomorphisms h_i with supports $(\{x : h_i(x) \neq x\})$ in U_i, which map each p_i into q_i, and are locally translations at p_i. The composition of the h_i gives the required homeomorphism h. The L_i and U_i can always be chosen to avoid the given finite set.

In the general case, where the arcs L_i are not necessarily disjoint, we may at least assume (by suitable small displacements of the L_i, if necessary) that the arcs L_i intersect in a finite subset F, and (by small reparameterization, if necessary) that for each fixed t in $[0, 1]$ the points $L_i(t)$, $i = 1, \ldots, N$, are distinct. Let $t_1 < t_2 < \cdots < t_k$ be all the values of t for which $L_i(t) \in F$ for some i. Choose numbers s_j for $j = 1, \ldots, k - 1$ so that

$$0 = s_0 \leq t_1 < s_1 < t_2 < \cdots < s_{k-1} < t_k < s_k = 1.$$

Next define $p_{ij} = L_i(s_j)$ and $q_{ij} = L_i(s_{j+1})$ for $i = 1, \ldots, N$ and $j = 0, \ldots, k - 1$. Then for each fixed j, the sets $\{p_{ij}\}_{i=1}^{N}$ and $\{q_{ij}\}_{i=1}^{N}$ satisfy the disjoint arc assumption (with respect to the arcs obtained by restricting the L_i to the interval $[s_j, s_{j+1}]$) and the distance condition $|p_{ij} - q_{ij}| < \epsilon(s_{j+1} - s_j)$. Hence by the special case already established, we obtain for each $j = 0, \ldots, k - 1$ a volume preserving homeomorphism h_j with $\|h_j\| < \epsilon(s_{j+1} - s_j)$ and $h_j(p_{ij}) = q_{ij}$. The composition $h = h_{k-1} \circ h_{k-2} \circ \cdots \circ h_0$ satisfies the requirements of the theorem.

Since all the constructions used in the above proof for I^n are local, they can be carried out on any manifold, and produce an h which is the composition of homeomorphisms supported by balls. Also note that the only property of the underlying (Euclidean) metric which was used in the proof was that two points at distance less than δ can be joined by an arc of length less than δ.   □

Theorem 2.4 is due to Oxtoby and Ulam [88]. The explicit construction given here comes from Alpern and Edwards [19].

3
Discrete Approximations

3.1 Introduction

Much of the early work in the 1970s on simplifying and extending the results of Oxtoby and Ulam was based on an observation of Peter Lax [80] that volume preserving homeomorphisms of the cube could be uniformly approximated by dyadic permutations. Lax's approximation technique, based on the combinatorial Marriage Theorem, was able to simply substitute for the original use of the Individual (Birkhoff) Ergodic Theorem by Oxtoby and Ulam. These discrete techniques were later replaced in the theory by more powerful methods (see Chapters 6 and 8). The results in this chapter and those that are based on it (Chapters 4, 5, and 7) will not be used elsewhere. However, the idea of approximating volume preserving homeomorphisms by dyadic permutations is very intuitive and has often led to the first proof of new results later improved on by the other methods.

We motivate the approach of this chapter by considering how volume preserving homeomorphisms may be approximately modeled on a digital computer. Suppose that h is a volume preserving homeomorphism of the space X which is the unit n-cube I^n or the n-torus T^n obtained from it by identifying opposite sides. We consider schemes by which we may input any point x in X, and obtain the point $h(x)$ as output. Of course we cannot actually input vectors of real numbers into a finite computer, nor can we expect to get them as output. More realistically we would input some number m of binary digits for each coordinate of x, and obtain similar output. Thus what the computer actually gives us is a map $h_c : \mathcal{D} \to \mathcal{D}$, where $\mathcal{D}$ is the set of all $N = 2^{mn}$ dyadic n-cubes of order m. (Such an n-cube is the product of intervals of the form $[k/2^m, (k+1)/2^m]$.) We now consider to what extent the approximate computer scheme h_c, which we require to be uniformly close to h, can model the properties of h. In particular, we ask the following questions:

(i) Since h is a bijection, can we always choose h_c to be a bijection, that is, a *permutation* of $\mathcal{D}$?

(ii) If h is ergodic (has no nontrivial invariant sets), can we choose h_c to be ergodic on $\mathcal{D}$, that is, a *cyclic permutation* of $\mathcal{D}$?

(iii) If X is the torus T^n it is easy to define for h (we shall do so formally below) the mean rotation of points around each of the fundamental circles of T^n. If the mean rotation for h is zero in each circle direction, can we choose h_c so that it is a *cyclic permutation which also has mean rotation vector zero*?

The first question was raised by Peter Lax [80], who used the Marriage Theorem of P. Hall to give an affirmative answer. The second question was answered affirmatively by the first author, who showed [6] that h_c can be taken to be cyclic *whether or not* h is ergodic. That the ergodicity of h is not required should not surprise us, given the result of Oxtoby and Ulam that any h can be uniformly approximated by an ergodic homeomorphism. However, we will turn this argument around in Chapter 7 by showing that the cyclic permutation approximation can be used to find a dense class of ergodic homeomorphisms. Finally, the affirmative answer to the last question was given by the authors [26], who used it to obtain a simple proof (presented in the next chapter) of Franks's result [62] that volume preserving homeomorphisms of T^n with mean rotation zero have a fixed point. So while the questions given above can all be motivated by purely computational considerations, their answers are not without theoretical consequences. The rest of this chapter is devoted to proving the affirmative answers to the three computational questions.

3.2 Dyadic Permutations

We now give the elegant proof of Lax on the approximation of volume preserving homeomorphisms by dyadic permutations. A dyadic permutation of I^n of order m is a map $P : I^n \to I^n$ which permutes the dyadic cubes of order m by simple translation. Its behavior on the boundary of these cubes will not matter, as P will only be considered as a member of the space $\mathcal{G}[I^n, \lambda]$, where zero measure behavior is irrelevant. The same definition applies to spaces like T^n which can be obtained from I^n by boundary identifications.

Theorem 3.1 (Lax) *Let X be any compact manifold obtained from I^n by boundary identifications, let h be a volume preserving homeomorphism of X, and let $\epsilon > 0$. Then for all sufficiently large m there is a dyadic permutation P of X of order m such that $\|P - h\| < \epsilon$.*

Proof Let $\mathcal{D} = \{\alpha_i,\ i = 1, \dots, N\}$ be the $N = 2^{mn}$ closed dyadic n-cubes of some large order m to be determined later. If we can find a dyadic permutation P of X satisfying

$$P(\alpha_i) \cap h(\alpha_i) \neq \emptyset, \quad i = 1, \dots, N \tag{3.1}$$

then for all $x \in X$ we will have that

$$|P(x) - h(x)| \leq |\alpha| + |h(\alpha)| < \epsilon, \text{ for sufficiently large } m. \tag{3.2}$$

Here $|h(\alpha)|$ denotes the maximum diameter of the sets $h(\alpha_i)$, and $|\alpha|$ denotes the diameter of each cube α_i. Say that cube α_i *knows* cube α_j if $\alpha_j \cap h(\alpha_i) \neq \emptyset$. The 'Marriage Theorem' of P. Hall [68] asserts that we can always assign a distinct cube among those it knows to each α_i, as long as the following condition is met: every set of any k cubes must together *know* at least k cubes. But since h preserves volume, any set A consisting of k cubes has an image $h(A)$ which also has the volume of k cubes, and hence must intersect at least k cubes. Thus Hall's condition is satisfied, and to all the cubes α_i we may assign distinct cubes $P(\alpha_i)$ satisfying (3.1) and hence (3.2). $\qquad\Box$

An alternative proof of Lax's result may be obtained by convexity considerations. Define the 'transition probability' matrix $\mathbf{T}$ of h with respect to the cubes α_i according to the formula

$$t_{ij} = \lambda(h(\alpha_i) \cap \alpha_j)/\lambda(\alpha_i).$$

The matrix $\mathbf{T}$ has all of its row and column sums equal to 1, a condition called *doubly stochastic*. Viewed as a subset of $N \times N$ dimensional Euclidean space, the doubly stochastic matrices form a closed convex set whose extreme points are the permutation matrices (0–1 matrices with a single 1 in each row and each column). The Birkhoff–von Neumann Theorem [42, 104] asserts that $\mathbf{T}$ can be represented as a convex combination of permutation matrices. The dyadic permutation P corresponding to any of the permutation matrices (p_{ij}) appearing in the representation of $\mathbf{T}$ (with positive coefficient a) will satisfy $\alpha_j = P(\alpha_i) \Leftrightarrow p_{ij} = 1 \Rightarrow t_{ij} > a \cdot 1 > 0$ and hence also condition (3.1).

It is worthwhile to compare Lax's result with an earlier result of Halmos [72, p. 65] asserting that the dyadic permutations are dense in $\mathcal{G}[I^n, \lambda]$ with respect to the weak topology. Lax's result has a finer topology (uniform) but only claims to approximate homeomorphisms.

3.3 Cyclic Dyadic Permutations

We now investigate whether the approximating dyadic permutation of Theorem 3.1 can be taken to be cyclic, that is, to permute the dyadic cubes in a single cycle. Given Theorem 3.1, this amounts to asking whether any dyadic permutation can be uniformly approximated, to within a fixed number of cube lengths, by a cyclic dyadic permutation. The answer is yes, and two cube lengths are sufficient. In other words, we can take P to be cyclic in (3.2) if we change the $|\alpha|$ to $3|\alpha|$. This result can be proved (see [6]) by an analysis which makes use of the adjacency structure of the set of dyadic n-cubes. However, we prefer to use a simpler and more general method based on the following combinatorial result from [7]:

Lemma 3.2 *Given any permutation ρ of $J = \{1, \ldots, N\}$, there is a cyclic permutation σ of J with $|\rho(j) - \sigma(j)| \leq 2$ for all j in J.*

Proof We describe an algorithm for constructing σ, given ρ. First assume that $N = 2m$ is even. Define permutations $r_1, r_2, \ldots, r_m$ recursively as follows: Let r_1 be either the identity or the transposition of 1 and 2, depending on whether those integers lie on the same cycle or different cycles of ρ, respectively. Observe that in either case they lie on the same cycle of $r_1 \circ \rho$. Recursively define r_j to be either the identity or the transposition of $2j - 1$ and $2j$, depending respectively on whether those integers lie on the same or different cycles of $r_{j-1} \circ r_{j-2} \circ \cdots \circ r_1 \circ \rho$. For brevity set $R \equiv r_m \circ \cdots \circ r_1$ and observe that under the permutation $R \circ \rho$ each consecutive odd–even pair of the form $2k-1, 2k$ lies on a single cycle. Furthermore since R is the composition of disjoint transpositions of consecutive integers we have that

$$|R(j) - j| \leq 1, \quad \text{for all } j \text{ in } J. \tag{3.3}$$

Similarly we define permutations $s_1, \ldots, s_{m-1}$ to link the even–odd pairs: Let s_j be either the identity or the transposition of $2j$ and $2j + 1$, depending respectively on whether those two numbers lie on the same or different cycles of the permutation $s_{j-1} \circ \cdots \circ s_1 \circ R \circ \rho$. Set $S \equiv$

$s_{m-1} \circ \cdots \circ s_1$ and $\sigma \equiv S \circ R \circ \rho$. By the same reasoning used for (3.3) we conclude that

$$|S(j) - j| \leq 1, \quad \text{for all } j \text{ in } J. \tag{3.4}$$

Combining (3.3) with (3.4) and the definition of σ gives

$$|\rho(j) - \sigma(j)| \leq 2, \quad \text{for all } j \text{ in } J.$$

To see that in fact σ is cyclic, observe that by construction every consecutive odd–even pair and every consecutive even–odd pair lies on a single cycle. It follows that there is only one cycle, completing the proof for N even. If $N = 2m + 1$ is odd the same algorithm, with an additional permutation s_m at the end, produces the required cyclic permutation.
□

To see if you have understood the algorithm given above, check that if you start with (in cycle notation) $\rho = [(1)\ (27)\ (3)\ (4)\ (56)\ (8)\ (9)]$ then you get $R = [(12)\ (34)\ (5)\ (6)\ (78)\ (9)]$, $S = [(1)\ (23)\ (45)\ (6)\ (7)\ (89)]$ and $\sigma = [(135642987)]$. Note that $\sigma \circ \rho^{-1} = [(13542)\ (6)\ (798)]$ is a permutation which moves no number more than two places. This simple combinatorial lemma may be thought of as a discrete version of the fact that 'ergodicity is a dense property'.

To apply the lemma to produce a cyclic dyadic approximation to the P of Theorem 3.1, simply number the $N = 2^{mn}$ dyadic cubes so that consecutive cubes share a common face. Then replace P by its cyclic approximation, which increases the error estimate (3.2) by at most $2\,|\alpha|$, proving the following cyclic version of Theorem 3.1.

Theorem 3.3 *Let X be any compact manifold obtained from I^n by boundary identifications, let h be a volume preserving homeomorphism of X, and let $\epsilon > 0$. Then for all sufficiently large m there is a cyclic dyadic permutation P of X, of order m, such that $\|P - h\| < \epsilon$.*

Of course there is nothing special about the *dyadic* cubes (of order m) obtained by dividing each axis of I^n in half m times. We could equally well apply the Marriage Theorem to k-adic cubes of order m, obtained by taking n-fold products of the form $[i/k^m, (i+1)/k^m]$, for any natural number k. So the above theorem applies as well to produce cyclic k-adic permutations. Sometimes we will need a version of the above theorem in which h is approximated by another volume preserving homeomorphism of X which agrees with P on the centers of the k-adic cubes. The result we will need is the following.

Corollary 3.4 *Let X be any compact manifold obtained from I^n by boundary identifications, let h be a volume preserving homeomorphism of X, let k be any natural number, and let $\epsilon > 0$. Then for all sufficiently large m, there is a volume preserving homeomorphism f of X, with $\|f\| < \epsilon$, such that fh cyclically permutes the centers of the $N = k^{mn}$ k-adic cubes of order m. Furthermore f can be chosen to fix any given finite set disjoint from these centers.*

Proof It follows from Theorem 3.3 that there is k-adic permutation P, of some k-adic decomposition $\{\sigma_j\}_{j=1}^N$ of order m, which satisfies $\|h - P\| < \epsilon$. Number the cubes $\{\sigma_j\}_{j=1}^N$ so that $P(\sigma_j) = \sigma_{j+1}$ (where $j+1$ is understood $\mathrm{mod}(N)$), and let p_j denote the center of the cube σ_j. Since $|h(p_j) - P(p_j)| < \epsilon$ for each $j = 1, \ldots, N$, it follows from Theorem 2.4 that there is a homeomorphism $f \in \mathcal{M}[I^n, \lambda]$ which fixes the given finite set, satisfies $\|f\| < \epsilon$, and is such that $f(h(p_j)) = P(p_j) = p_{j+1}$ for all $j = 1, \ldots, N$. It follows that the homeomorphism fh cyclically permutes the centers p_j. $\square$

The techniques used in this subsection can be extended to approximate any permutation of the vertices of a connected graph by a cyclic permutation of these vertices such that the two permutations differ by at most 6 edges [9]. (Lemma 3.2 gives a smaller error of 2 edges for the line graph.)

3.4 Rotationless Dyadic Permutations

We now restrict our attention to the case where the underlying manifold X is the n-dimensional torus T^n or the annulus (or cylinder) $A = S^1 \times I$. It is convenient to assume that the distance is given by the maximum metric $|x - y| = \max_i |x_i - y_i|$, with length $|x| = \max_i |x_i|$, and that $\|g\|$ denotes the corresponding maximum norm of an automorphism g. Let $\hat{X}$ be the universal covering space of X, that is, R^n (Euclidean n-space) or $R^1 \times I$ (the strip). We consider question (iii) of the introduction which asked whether the approximation h_c to a homeomorphism h with mean rotation zero could be chosen to also have mean rotation zero. We call a transformation with mean rotation zero *rotationless*.

Define the *mean rotation vector* v of a volume preserving homeomorphism h of X (or the mean translation $v_t(\hat{h})$ of a lift $\hat{h}$) by the formula

$$v(h) = v_t(\hat{h}) = \int_\Omega (\hat{h}(x) - x) \, d\lambda(x), \qquad (3.5)$$

where $\hat{h}$ is a lift of h to the covering space $\hat{X}$, $\Omega = I \times I$ is a fundamental domain and λ, as usual, denotes Lebesgue measure. We call (3.5) the mean translation of $\hat{h}$. Since $\hat{h}$ is defined only up to unit translations in each direction, v is defined modulo 1 in each coordinate. Observe that according to this definition the rotation $x \mapsto x + w$ (taken mod 1 in each coordinate) will have mean rotation w. Let f and g be two homeomorphisms of the torus (or the annulus). By choosing the lift $\widehat{fg}$ appropriately, and noting that $\widehat{fg}(x) - x = \widehat{fg}(x) - \hat{g}(x) + \hat{g}(x) - x$, it is easy to see that the rotation vector of the lift $\widehat{fg}$ satisfies

$$v_t(\widehat{fg}) = v_t(\hat{f}) + v_t(\hat{g}). \qquad (3.6)$$

Consequently, the mean rotation vector of the composition of f and g satisfies $v(fg) = v(f) + v(g)$. When X is the annulus A, the mean rotation vector will always have the second coordinate zero.

We will find it useful to extend the definition of mean rotation to volume preserving automorphisms $g \in \mathcal{G}[X, \lambda]$ so that we can in particular apply the notion to dyadic permutations. If such a g satisfies $\|g - h\| < 1/2$ for some homeomorphism h, then there is clearly a unique lift $\hat{g}$ to $\hat{X}$ (that is, with $\pi\hat{g} = g\pi$, where π denotes the covering projection) satisfying $\|\hat{g} - \hat{h}\| < 1$. As long as there is such a homeomorphism near g, we can use this lift $\hat{g}$ to define $v(g)$ as in (3.5) (otherwise $v(g)$ might not be well defined). It is easy to see that if P is a dyadic permutation (near some homeomorphism) with well defined mean rotation, each coordinate of its mean rotation vector is an integer multiple of the volume $(1/N)$ of the cubes it permutes. A final general observation is that $v(\text{identity}) = \vec{0}$ and that for any $g \in \mathcal{G}[X, \lambda]$, $|v(g)| \leq \|g\|$. To explain the last inequality in say the first coordinate, it is enough to observe that the average amount moved to the right is bounded by the maximum that any point moves to the right. Of course this last inequality shows that if h is uniformly approximated by h_c (that is, the dyadic permutation P of Theorem 3.1), then $v(h_c) \equiv v(P)$ will be close to $v(h)$. The only problem to be dealt with in the following theorem is thus how to ensure exact equality (both zero).

Theorem 3.5 *Let X be the n-torus T^n or the annulus A, let h be a volume preserving homeomorphism of X with mean rotation zero, and* let $\epsilon > 0$. *Then for all sufficiently large m there is an mth order dyadic permutation Q of X such that $\|Q - h\| < \epsilon$, and the mean rotation of Q is zero.*

Proof We consider only the case $X = T^n$, as the proof for the annulus is essentially the same, but easier. According to Theorem 3.1 there is an mth order dyadic permutation P with $\|P - h\| < \epsilon/3$. Let $M = 2^m > 3/\epsilon$ denote the number of dyadic cubes in each coordinate direction, so that the diameter of each cube (in the max metric) is $1/M$. Since we are assuming that $v(h) = \vec{0}$, it follows that

$$|v(P)| = |v(P) - v(h)| \le \|P - h\| < \epsilon/3.$$

To obtain a mean rotation zero dyadic permutation Q near P (and hence near h) we simply counteract any mean rotation in some coordinate by a rotation (or as close to a rotation as we can get dyadically) in the opposite direction. We begin with the first coordinate $v_1(P)$, which we may assume without loss of generality is positive. In particular, we may write $v_1(P) = k/N$, for some integer k between 1 and $N - 1$, where $N = M^n$ is the number of dyadic cubes in X. Hence we may write $v_1(P) = \left(aM^{n-1} + b\right)/M^n$ for some integers a and b with $0 \le a \le M-1$ and $0 \le b < M^{n-1}$. It follows that

$$\frac{a}{M} \le v_1(P) \le \epsilon/3.$$

We consider the $N = M^n$ dyadic cubes of T^n as being arranged in M^{n-1} horizontal strips (cylinders) each consisting of M dyadic cubes. Let R_1 denote the dyadic permutation which shifts (mod 1) all of these rows a cubes to the left, and b of these rows an additional cube to the left. Then

$$v_1\left(R_1\right) = -\left(a + b/M^{n-1}\right)\left(1/M\right) = -v_1(P)$$

so that by (3.6), $v_1(R_1 \circ P) = 0$. Note that since R_1 moves no point more than $a+1$ cube lengths, we have that $\|R_1\| < (a + 1)/M \le \epsilon/3 + 1/M < 2\epsilon/3$. Perform a similar rotational correction R_j in the jth coordinate, and let R be their product. Since we are using the maximum metric, we have $\|R\| \le 2\epsilon/3$. Then the dyadic permutation $Q = R \circ P$ satisfies $v(Q) = \vec{0}$, $\|P - Q\| = \|R\| \le 2\epsilon/3$, and hence $\|Q - h\| < \epsilon$. $\qquad\square$

The argument used in the previous section to show that the permutation P of Theorem 3.1 could be taken to be cyclic can also be used to show that the permutation Q of Theorem 3.6 can be taken to be cyclic. Since that argument numbered the dyadic cubes so that consecutive cubes shared a common *interior* face of I^n the property of having mean rotation zero will not be destroyed in the process. Also note that if a cyclic dyadic permutation Q has mean rotation zero, then the lift $\hat{Q}$ with $\int_\Omega (\hat{Q}(x) - x) \, d\lambda(x) = \vec{0}$ to a dyadic permutation of the covering space consists of identical cycles of the same length N. (Otherwise we would have that $\hat{Q}^N(\alpha) = \alpha + w$ for all α and some vector $w \neq \vec{0}$, in which case $v(Q)$ would be equal to $w/N \neq \vec{0}$.) Hence we have proved the following.

Corollary 3.6 *Let X be the n-torus T^n or the annulus A, let h be a mean rotation zero volume preserving homeomorphism of X, and let $\hat{h}$ be a lift to the covering space R^n or $R^1 \times I$ satisfying*

$$\int_\Omega \left(\hat{h}(x) - x \right) d\lambda(x) = \vec{0}.$$

Then for all $\epsilon > 0$ and sufficiently large m there is an mth order cyclic dyadic permutation Q of X with a lift $\hat{Q}$ which permutes all the mth order dyadic cubes of the covering space in cycles of length 2^{mn} and satisfies $\|\hat{Q} - \hat{h}\| < \epsilon$.

For applications to the fixed point theorems of the next chapter, we will need the following consequence of this result.

Theorem 3.7 *Any volume preserving homeomorphism of the n-torus or annulus X which has mean rotation zero can be uniformly approximated by a similar homeomorphism which has a lift with a periodic point.*

Proof Using the proof and notation of the previous corollary, let p_i, $i = 1, \ldots, 2^{mn}$, be the centers of the permuted cubes, numbered cyclically so that $Q(p_i) = p_{i+1} \pmod{2^{mn}}$. Since $\|Q(p_i) - h(p_i)\| < \epsilon$, it follows from Theorem 2.4 that there is a volume preserving homeomorphism f of X, with $\|f\| < \epsilon$, such that $f(h(p_i)) = p_{i+1}$(index i mod 2^{mn}). Hence the lift of fh which is near $\hat{h}$ permutes all the centers of the mth order dyadic cubes of the covering space in cycles of length 2^{mn}. Since $\|fh - h\| < \epsilon$, and ϵ was arbitrary, this completes the proof. $\qquad\square$

4

Transitive Homeomorphisms of I^n and R^n

4.1 Transitive Homeomorphisms

A mapping h of a topological space is called *transitive* if for any open sets U and V, there is a positive integer k with $h^k U \cap V \neq \emptyset$. For manifolds (actually, for any complete, separable metric space without isolated points) this is equivalent to the existence of a dense orbit under h. For some spaces, it is easy to exhibit transitive homeomorphisms (e.g., irrational rotations of the circle), while for other spaces where they exist more subtlety is required. Besicovitch [40] first defined such a transformation for the plane in 1937, partially answering a question of Ulam (see Chapter 12). In the same year, Oxtoby [89] used the Baire Category Theorem to demonstrate that in fact transitivity is typical for volume preserving homeomorphisms of the cube. At that time, the existence of such transformations had not been established. Recently Xu ([108, 109]) has shown how Besicovitch's transitive homeomorphism of the plane can be used to construct an explicit transitive homeomorphism of the closed unit square and Cairns, Jessup and Nicolau [48] have given examples on the 2-sphere and more generally on quotients of tori.

In this chapter we present a simplification of Oxtoby's original proof based on the combinatorial techniques of Chapter 3, and then we extend this method to show the existence of spatially periodic transitive homeomorphisms of R^n, or equivalently, rotationless homeomorphisms of the torus with transitive lifts. The first result (Theorem 4.1) is a weaker version of the similar result for ergodic homeomorphisms (Theorem 7.1) to be given later (independently) and is presented here mainly to give the reader an early introduction to the category techniques to be used throughout the book. The second result (Theorem 4.2) will however not be improved upon here, as the corresponding result for ergodic homeomorphisms is an open question. A slightly stronger property than transitivity called topological weak mixing is shown to hold typically in the volume preserving homeomorphisms in the next section. We conclude this chapter by showing that a modification of these constructions can be used to prove the existence of a volume preserving *chaotic* homeomorphism of I^n ($n \geq 2$). The notion of chaos used here is based on a weaker concept due to Devaney [56].

The results of this chapter will not be used in later chapters.

4.2 A Transitive Homeomorphism of I^n

To prove the existence of a transitive homeomorphism of the cube by category methods, we must impose the *additional* requirement of volume preservation. To see this, simply observe that if a homeomorphism maps some closed set into its interior, so does any sufficiently close homeomorphism, which therefore cannot be transitive. Thus transitive homeomorphisms are not even dense in the space of all homeomorphisms of the cube. The problem caused by a closed set being mapped into its interior is avoided by assuming volume preservation, and in fact this assumption (restricting to the space $\mathcal{M}[I^n, \lambda]$) is sufficient to make transitivity typical.

Theorem 4.1 *The subset $\mathcal{T}$ of $\mathcal{M}[I^n, \lambda]$ consisting of transitive volume preserving homeomorphisms of I^n contains a dense G_δ subset of $\mathcal{M}[I^n, \lambda]$ with respect to the uniform topology.*

Proof Let τ_i, $i = 1, \ldots, \infty$, be an enumeration of all the open dyadic cubes of I^n, of all orders. Let

$$\mathcal{T}_{ij} = \left\{ h \in \mathcal{M}[I^n, \lambda] : \exists k \geq 1, \ h^k(\tau_i) \cap \tau_j \neq \emptyset \right\}.$$

These sets are unions of open sets, hence open. Since every open set in I^n contains a dyadic cube, it follows from the definition of transitivity that $\bigcap_{ij} \mathcal{T}_{ij} \subset \mathcal{T}$. According to the Baire Category Theorem (see Theorem 2.1), it remains only to establish that each set $\mathcal{T}_{ij}$ is dense in $\mathcal{M}[I^n, \lambda]$. So let $h \in \mathcal{M}[I^n, \lambda]$ and $\epsilon > 0$ be given. It follows from Theorem 3.3 that there is a cyclic dyadic permutation P, of some decomposition $\{\sigma_m\}_{m=1}^N$ refining the decompositions τ_i and τ_j, which satisfies $\|h - P\| < \epsilon$. Number the cubes $\{\sigma_m\}_{m=1}^N$ so that $P(\sigma_m) = \sigma_{m+1}$ where arithmetic on the indices is done mod N, and let p_m denote the center of the cube σ_m. Since $|h(p_m) - P(p_m)| < \epsilon$ for each $m = 1, \ldots, N$, it follows from

Theorem 2.4 that there is a homeomorphism $f \in \mathcal{M}[I^n, \lambda]$ which is the identity on the boundary, with $\|f\| < \epsilon$, such that $f(h(p_m)) = P(p_m) = p_{m+1}$ for all $m = 1, \ldots, N$. It follows that the homeomorphism fh cyclically permutes the centers p_m and hence belongs to the set $\mathcal{T}_{ij}$, since both τ_i and τ_j contain at least one of these centers. Furthermore $\|fh - h\| = \|f\| < \epsilon$, as required. $\qquad\Box$

A slightly earlier version of this proof, adapted from [5], is included in the appendix to the second edition of Oxtoby's book [91].

4.3 A Transitive Homeomorphism of R^n

If a torus homeomorphism has a transitive lift to the covering space R^n then it is easy to see that it must be transitive itself (note that such a lift of a torus homeomorphism to R^n is spatially periodic). However, the converse of this statement is obviously false, so that the techniques of the previous section must be improved to obtain a torus homeomorphism with a transitive lift.

Theorem 4.2 *Let $\mathcal{M}^0[T^n, \lambda]$ denote the space of all homeomorphisms of the n-dimensional torus T^n, $n \geq 2$, which have mean rotation zero. Then the subset $\mathcal{TL}$ of $\mathcal{M}^0[T^n, \lambda]$, consisting of homeomorphisms whose mean translation zero lift to R^n is transitive, contains a dense G_δ subset.*

Proof The proof begins almost identically to the previous one. Let τ_i, $i = 1, \ldots, \infty$, be an enumeration of all the open dyadic cubes of R^n, of all orders. Define the sets

$$\mathcal{T}_{ij} = \left\{ h \in \mathcal{M}^0[T^n, \lambda] : \exists k \geq 1, \ \hat{h}^k(\tau_i) \cap \tau_j \neq \emptyset \right\},$$

where the $\hat{\ }$ denotes the unique lift to R^n with mean translation zero. As in the previous proof, we have that $\bigcap_{ij} \mathcal{T}_{ij} \subset \mathcal{TL}$. According to the Baire Category Theorem, it remains only to establish that each set $\mathcal{T}_{ij}$ is dense in $\mathcal{M}^0[T^n, \lambda]$. So given an $h \in \mathcal{M}^0[T^n, \lambda]$ and $\epsilon > 0$, we must find an $f \in \mathcal{M}^0[T^n, \lambda]$ with $\|f\| < \epsilon$ and $fh \in \mathcal{T}_{ij}$.

We now describe the construction of the required homeomorphism f. By Corollary 3.6, there is a cyclic dyadic permutation P, of some decomposition $\{\sigma_m\}_{m=1}^N$ of T^n with diameter less than $\epsilon/2$ and of higher order than that of τ_i and τ_j, which satisfies $\|h - P\| < \epsilon/2$. Furthermore, P has a lift $\hat{P}$ with $\|\hat{h} - \hat{P}\| < \epsilon/2$, which permutes the R^n decomposition $\sigma_{s,m}$, $m = 1, \ldots, N$, $s = 1, \ldots, \infty$, via the formula $\hat{P}(\sigma_{s,m}) = \sigma_{s',m+1}$

(where arithmetic on m is modulo N and $\sigma_{s,m}$ projects onto σ_m via the covering projection $\pi : R^n \to T^n$).

We now select a sequence of points $q_k \in R^n$, $k = 1, \ldots, K$, of some length K, with the following properties:

- $|q_{k+1} - q_k| < \epsilon/2$, for $k = 1, \ldots, K - 1$
- $q_1 \in \tau_i$, and $q_K \in \tau_j$
- The projections $\pi(q_k)$, $k = 1, \ldots, K$, are distinct points of the torus T^n, each lying in distinct relative positions in the interior of some cube σ. (The last requirement ensures that the points $P^m(\pi(q_k))$, $m = 0, \ldots, N - 1$; $k = 1, \ldots, K$, are all distinct.)

Define $q_{k,m} = \hat{P}^m(q_k)$, for $k = 1, \ldots, K$, and $m = 0, \ldots, N - 1$. Observe that for each fixed k, $\hat{P}$ permutes the points $q_{k,m}$ in a cycle of length N. To link all K of these N-cycles into a single chain, going from $q_1 = q_{1,0}$ in τ_i to $q_K = q_{K,0}$ in τ_j, we now use Theorem 2.4 to define a homeomorphism $f \in \mathcal{M}^0[T^n, \lambda]$ mapping $h(\pi(q_{k,m}))$ into $\pi(q_{k,m+1})$ for $m < N-1$ and $h(\pi(q_{k,N-1}))$ into $\pi(q_{k+1,0}) = \pi(q_{k+1})$. Since $|h\pi(q_{k,m}) - \pi(q_{k,m+1})| = |h\pi(q_{k,m}) - P\pi(q_{k,m})| < \epsilon/2$ for all m and $|h\pi(q_{k,N-1}) - \pi(q_{k+1})| \leq |h\pi(q_{k,N-1}) - P\pi(q_{k,N-1})| + |P\pi(q_{k,N-1}) - \pi(q_{k+1})| < \epsilon/2 + \epsilon/2 = \epsilon$ (since $\hat{P}(q_{k,N-1}) = q_{k,0} = q_k$), we may choose f with $\|f\| < \epsilon$. Observe that $(\widehat{fh})^{NK}(q_1) = q_K$. Also, by the last part of Theorem 2.4, f can be chosen to be the composition of homeomorphisms supported by balls. Since a homeomorphism of T^n with support in a small ball is rotationless, it follows from equation (3.6) that a composition of such homeomorphisms is also rotationless. Thus we can choose f so that it is rotationless. Therefore $fh \in \mathcal{T}_{ij}$, and the proof is completed. $\qquad\square$

Related work can be found in [25, 28, 29, 30].

4.4 Topological Weak Mixing

Earlier in this chapter we showed that transitivity is typical for volume preserving homeomorphisms of I^n. We now show how the same technique can be used to establish typicality for the stronger property called *topological weak mixing*, defined below.

Definition 4.3 *A map* $f : X \to X$ *of a topological space* X *is called topologically weak mixing if given any nonempty open sets* U_1, V_1, U_2, V_2, *there is a common iterate* i *such that both* $f^i(U_1) \cap V_1$ *and* $f^i(U_2) \cap V_2$ *are nonempty.*

It is worth observing that if f is topologically weak mixing then it is transitive. In order to show that such transformations are dense in $\mathcal{M}[I^n, \lambda]$, we will need the following result.

Lemma 4.4 *Let h be a volume preserving homeomorphism of I^n, let $\epsilon > 0$, and let U_1, V_1, U_2, V_2, be given nonempty open sets. Then there is another volume preserving homeomorphism f with $\|f\| < \epsilon$, such that fh has two cyclic orbits of relatively prime lengths, one of which enters U_1 and V_1, the other U_2 and V_2. Furthermore we can choose f so as to fix any given finite set. The lengths of the two orbits can be respectively chosen from the powers of any two relatively prime numbers.*

Proof Apply Corollary 3.4 to produce $f_1 \in \mathcal{M}[I^n, \lambda]$ with $\|f_1\| < \epsilon/2$ such that $f_1 h$ cyclically permutes the centers of some dyadic decomposition sufficiently fine so that it has cubes contained in U_1 and V_1 whose centers are not in the given finite set. Once f_1 is chosen, pick a $\delta > 0$ sufficiently small so that $\|f_2\| < \delta$ implies $\|f_2 f_1\| < \epsilon$. Apply Corollary 3.4 again to produce an $f_2 \in \mathcal{M}[I^n, \lambda]$ which fixes the centers permuted by $f_1 h$, satisfies $\|f_2\| < \delta$, and is such that $f_2 f_1 h = fh$ cyclically permutes the centers of some triadic (3-adic) decomposition containing cubes in U_2 and V_2. Thus fh has a cyclic orbit with length some power of 2 entering U_1 and V_1 and another with length a power of 3 entering U_2 and V_2. □

We can now prove that topological weak mixing is typical. The proof is similar to our earlier result for transitivity, so we will emphasize only the differences.

Theorem 4.5 *The subset $\mathcal{W}$ of $\mathcal{M}[I^n, \lambda]$, consisting of all topologically weak mixing volume preserving homeomorphisms of I^n, contains a dense G_δ subset of $\mathcal{M}[I^n, \lambda]$ with respect to the uniform topology.*

Proof Fix some countable base for the open subsets of I^n, and let $(U_1^r, V_1^r, U_2^r, V_2^r)$, $r = 1, 2, \ldots$ be an enumeration of all 4-tuples of such nonempty open sets. Define the sets $\mathcal{T}_r$, $r = 1, 2, \ldots$, by

$$\mathcal{T}_r = \{g \in \mathcal{M}[I^n, \lambda] : \exists i \geq 1 : g^i U_j^r \cap V_j^r \neq \emptyset, \ j = 1, 2\}.$$

Since each set $\mathcal{T}_r$ is open, and $\bigcap_r \mathcal{T}_r \subset \mathcal{W}$, we need only prove that each set $\mathcal{T}_r$ is dense in $\mathcal{M}[I^n, \lambda]$. We claim that given $h \in \mathcal{M}[I^n, \lambda]$ and $\epsilon > 0$, the approximation fh given by the previous lemma (with appropriate superscripts) belongs to $\mathcal{T}_r$. Since there are cyclic fh-orbits of relatively prime lengths N_j which enter U_j^r and V_j^r, $j = 1, 2$, there are points $p_j \in U_j^r$ with

$$(fh)^i p_1 \in V_1^r \text{ for } i = a \,(\mathrm{mod}\, N_1), \text{and}$$
$$(fh)^i p_2 \in V_2^r \text{ for } i = b \,(\mathrm{mod}\, N_2).$$

Since N_1 and N_2 are relatively prime, there are infinitely many positive integers i satisfying both equations and for these i we have $(fh)^i U_j^r \cap V_j^r \neq \emptyset$, $j = 1, 2$. Hence fh belongs to $\mathcal{T}_r$, as claimed. $\qquad\square$

4.5 A Chaotic Homeomorphism of I^n

In this section we modify the earlier constructions of this chapter to produce a volume preserving homeomorphism of I^n which is chaotic. Our notion of chaos is based on a weaker concept due to Devaney [56]. See also related work in [29, 30].

Definition 4.6 *A map $h : X \to X$ of a metric space (X, d) is called* Devaney-chaotic *if it satisfies the following three conditions:*

(i) *h is transitive*

(ii) *the periodic points of h are dense in X*

(iii) *h has 'sensitive dependence on initial conditions'. This means that for some $D > 0$, for any $x \in X$ and $\epsilon > 0$ there is always some y with $d(x, y) < \epsilon$ and some $i \geq 1$ with $d(h^i x, h^i y) \geq D$.*

It is known [37] that in fact the first two conditions (i) and (ii) imply the third, but we prefer to keep the definition in this form for the purposes of later comparison. We define a stronger notion of chaos, which we call *maximal chaos*, by strengthening the notion of sensitive dependence, as follows.

Definition 4.7 *A map $h : X \to X$ of a metric space (X, d) is called* maximally chaotic *if it satisfies the following three conditions:*

(i) *h is transitive*

(ii) *the periodic points of h are dense in X*

(iii) *h has 'maximal dependence on initial conditions'. This means that for any nonempty open subset U of X, $\overline{\lim}_i \, d\left(h^i U\right) = d(X)$, where $d(A)$ denotes the diameter of A.*

Condition (iii) is stronger than sensitive dependence on initial conditions and in fact implies that condition with any value of D less than $d(X)/2$. This is maximal because we might have that x is a fixed point at the metric center of X. The following result extends a similar result of Aarts and Daalderop [2] for Devaney-chaotic volume preserving homeomorphisms (see also Alpern [17] for compact manifolds and Alpern and Prasad [30] for the noncompact case).

Theorem 4.8 *Given any $h \in \mathcal{M}[I^n, \lambda]$, and $\epsilon > 0$, there is an $f \in \mathcal{M}[I^n, \lambda]$ of norm $\|f\| < \epsilon$ for which fh is maximally chaotic.*

Proof As in the construction of topologically weak mixing homeomorphisms, we fix some countable base for the open subsets of I^n, and let $(U_1^r, V_1^r, U_2^r, V_2^r)$, $r = 1, 2, \ldots$ be an enumeration of all 4-tuples of such nonempty open sets. Apply Lemma 4.4 to obtain $f_1 \in \mathcal{M}[I^n, \lambda]$ with $\|f_1\| < \epsilon/2$ such that $f_1 h$ has periodic orbits O_j^1, $j = 1, 2$, which respectively enter both U_j^1 and V_j^1, and the length of O_j^1 is a power of $j + 1$. Suppose we have constructed automorphisms $f_i \in \mathcal{M}[I^n, \lambda]$, $i = 1, \ldots, r$, such that the uniform distance $u(f_r f_{r-1} \ldots f_1, identity) < \epsilon/2 + \epsilon/4 + \cdots + \epsilon/2^r$ and $f_r f_{r-1} \ldots f_1 h$ has periodic orbits O_j^i, $i = 1, \ldots, r$ and $j = 1, 2$, which have length some power of $j + 1$ and enter both sets U_j^r and V_j^r. Define $F_r = f_r f_{r-1} \ldots f_1$. Choose δ so that if $\|g\| < \delta$ then $u(gF_r, F_r) < \epsilon/2^{r+1}$. Apply Lemma 4.4 again to obtain $f_{r+1} \in \mathcal{M}[I^n, \lambda]$ with $\|f_{r+1}\| < \delta$, which fixes all the orbits O_j^i, $i = 1, \ldots, r$ and $j = 1, 2$ and is such that $f_{r+1} f_r \ldots f_1 h$ has periodic orbits O_j^{r+1}, $j = 1, 2$, which have length some power of $j + 1$ and enter both sets U_j^{r+1} and V_j^{r+1}. Since $\|f_{r+1}\| < \delta$, $u(F_{r+1}, F_r) < \epsilon/2^{r+1}$. Hence F_r is Cauchy with respect to the uniform metric u. Thus $\lim_{r \to \infty} f_r f_{r-1} \ldots f_1$ converges to some $f \in \mathcal{M}[I^n, \lambda]$, with $\|f\| < \epsilon$ such that fh has periodic orbits O_j^i, $i = 1, 2, \ldots$ and $j = 1, 2$, which have length some power of $j + 1$ and enter both sets U_j^i and V_j^i. We claim that fh is maximally chaotic. Since the orbits O_j^i enter all of a basic family of open sets, the periodic points of fh are dense. Since fh belongs to each set $\mathcal{T}_r$ in the previous theorem, it is topologically weak mixing, and hence transitive. Finally, we show that fh (or any topologically weak mixing homeomorphism) satisfies the third condition regarding maximal dependence on initial conditions. If $D < d(X)$, we can find basic open sets V_1^r and V_2^r with $d(y_1, y_2) > D$ for any $y_1 \in V_1^r$, $y_2 \in V_2^r$. Choose U_1^r and U_2^r contained within the open set U given in the definition of maximal dependence on initial conditions. Using the same idea as in the proof of the previous theorem, there are infinitely many positive integers i for which $(fh)^i U_j^r \cap V_j^r \neq \emptyset$, $j = 1, 2$, and hence for which $d((fh)^i U) > D$. $\qquad\square$

Very recently, Daalderop and Fokkink [55] and also Akin [4] have shown that property (ii) of Definitions 4.6 or 4.7 (having dense periodic points) is a residual property in $\mathcal{M}[I^n, \lambda]$. If that result is combined with Theorem 4.8, then we see that maximal chaos is also a residual property in $\mathcal{M}[I^n, \lambda]$. These results apply more generally to compact manifolds.

If volume preservation is not required, then the main results for existence of chaotic actions of compact manifolds are given in [49] and [47].

4.6 Periodic Approximations

We conclude this chapter with a construction (originally due to Oxtoby and Ulam and generalized by the first author) that approximates an automorphism of the cube by another which is locally linear almost everywhere. In fact, a degenerate case of this construction gives uniform approximation by Devaney-chaotic automorphisms on arbitrary compact manifolds. Of course this latter application was not realized when the construction was formulated.

For any α, $0 \leq \alpha \leq 1$, define an α-cube of order m to be a closed cube concentric to and with a side length α times that of some dyadic cube of order m. (A 0-cube is simply the center of a dyadic cube.) These fractional cubes will be used later in Lemma 6.6. We call α a *dyadic rational* if it has the form $k/2^m$. The following result was proved in the first author's thesis [5] using successive application of our Corollary 3.4 (to permute the centers of the α-cubes) and the Annulus Theorem, our Theorem 12.1 (to enlarge these centers to α-cubes). Actually only the case $\alpha = 1/2$ was stated there, but the proof is the same (and can be found in [17] for all α in $[0, 1)$). It generalizes an earlier result of Oxtoby and Ulam [88, Theorem 12] which applied only to the unit cube itself and to automorphisms h which are isotopic to the identity on the boundary of the unit cube.

Theorem 4.9 *Let X be a compact manifold obtained by making boundary identifications of I^n, let α, $0 \leq \alpha < 1$, be any dyadic rational and let $\epsilon > 0$. Then given any automorphism h of X, there is an automor-*

phism f of X with $\|f\| < \epsilon$, and an infinite strictly increasing sequence of positive integers $m_1, m_2, \ldots$ with the following property: For each $k = 1, 2, \ldots$, the α-cubes of order m_k which are not contained in any α-cube of order $m_1, m_2, \ldots, m_{k-1}$ are rigidly permuted in a single cycle by $\bar{h} = fh$.

The consequences of the above result are very different in the two cases $\alpha > 0$ and $\alpha = 0$. For $\alpha > 0$, the permuted α-cubes have total volume 1. This means that for almost every point x in I^n, x is periodic under $\bar{h}$, and in fact belongs to a closed cube on which $\bar{h}$ acts rigidly and hence linearly. So we immediately obtain the following corollary, which we state for simplicity on the cube.

Corollary 4.10 *A volume preserving homeomorphism h of the closed unit cube $I^n, n \geq 2$, can be uniformly approximated by another which is measure theoretically periodic and is locally linear almost everywhere.*

For the case $\alpha = 0$, the α-cubes are simply the centers of their dyadic cubes, and so *all* these centers of cubes of order m_k are permuted in a single cycle of $\bar{h}$. It follows that since any given pair U, V of nonempty open subsets of I^n contain dyadic cubes of some order m_k, there is an orbit of $\bar{h}$ which enters both U and V. This condition clearly implies that $\bar{h}$ (i) is transitive, and (ii) has a dense set of periodic points. Consequently we have another proof that

Corollary 4.11 *Every volume preserving homeomorphism of the closed unit n-cube, $n \geq 2$, with boundary identifications, can be uniformly approximated by another which is chaotic in the sense of Devaney.*

5

Fixed Points and Area Preservation

5.1 Introduction

Fixed point theorems are usually purely topological in nature, and do not usually have any measure theoretic hypotheses. However, there are three surfaces where the assumption that a homeomorphism is area preserving, by itself or with additional assumptions, implies the existence of a fixed point: the open square, the torus, and the annulus. The reason only 2-dimensional manifolds are covered is that all these results follow from a purely topological fixed point theorem of Brouwer for homeomorphisms of the plane, known as the 'Plane Translation Theorem'. This theorem says that if an orientation preserving homeomorphism of the plane has no fixed point then it is 'like a translation'. This phrase can be made precise in various ways, but it will be sufficient for our purposes here to take it to mean 'has no periodic points'.

Since the issue of fixed points is not a main concern of this book, we will not attempt to give the strongest forms of theorems, but merely show how results obtained earlier in the book can give simple demonstrations of the existence of fixed points. References to the stronger results of Franks and Flucher will be given.

The organization of this chapter is as follows. In Section 5.2 we state a special case of Brouwer's Plane Translation Theorem due to Andrea [32]. We apply this in Section 5.3 to prove a result of Montgomery [86] that any orientation preserving, area preserving homeomorphism of the open square has a fixed point. In Section 5.4 we use the techniques of the last chapter to prove Franks's result [62] that any area preserving homeomorphism of the torus which is homotopic to the identity and has mean rotation zero has a fixed point. This is a topological version of a well known theorem of Conley and Zehnder [53] for diffeomorphisms.

31

Finally, we show in Section 5.5 that the approach of Section 5.4 also yields a proof of Poincaré's 'Last Geometric Theorem' [94], that any orientation preserving, area preserving homeomorphism of the annulus which rotates the two bounding circles in opposite directions has a fixed point. We will mention extensions to these results which give additional fixed points.

5.2 The Plane Translation Theorem

Brouwer's 1912 paper [44] was a deep investigation into the properties of orientation preserving homeomorphisms of the plane which have no fixed points. He found that such homeomorphisms possess many properties of nontrivial translations $h_v(x) = x + v$, where v is not the zero vector. For example, consider the set F of points lying between a line L_1 perpendicular to v and its image $L_2 = h_v(L_1)$, including those points on L_2. Such a set F is called a *translation field* for h_v in that its iterates under positive and negative powers of h_v fill up the plane without overlapping. Brouwer showed that such a set (lying between two homeomorphs of the real line) always exists for an orientation preserving homeomorphism of the plane without fixed points. Another version of this type of property is given in (5.1). Note that if a translation has no fixed points then it is nontrivial and also has no periodic points. A special case of the Brouwer Plane Translation Theorem, as observed by Andrea [32], says that this property of translations is possessed by fixed point free homeomorphisms of the plane.

Theorem 5.1 (Plane Translation Theorem) *Suppose that h is any orientation preserving homeomorphism of the plane which has no fixed point. Then it satisfies the following property:*

$$\begin{array}{ll} \text{If } C \text{ is compact and connected and } h(C) \cap C = \emptyset, \\ \text{then } h^m(C) \cap C = \emptyset, \text{ for all integers } m \neq 0, \end{array} \quad (5.1)$$

and in particular, h has no periodic points. $\quad (5.2)$

Since this is a purely topological result whose proof uses techniques which are not otherwise useful in the book, we will not present a proof. Those interested can find a nice proof of (5.2) in [59] and a short argument that (5.2) implies (5.1) in [32].

To apply the above result in a measure theoretic context, we need the following definition and application of the notion of μ-recurrence.

Definition 5.2 *An automorphism g of a measure space (X, Σ, μ) is said to be recurrent, or μ-recurrent, if for any $A \in \Sigma$ with $\mu(A) > 0$, μ-almost every point of A eventually returns to A under some positive iterate of g. That is, $\mu\left(A - \bigcup_{m \geq 1} g^{-m}(A)\right) = 0$.*

Corollary 5.3 *A sufficient condition that an orientation preserving homeomorphism of the plane has a fixed point is that it is μ-recurrent for some invariant measure μ which is positive on open sets.*

Proof Suppose some such homeomorphism h has no fixed point. Then we can find a closed disk C of positive radius with $h(C) \cap C = \emptyset$, which by (5.1) satisfies $h^m(C) \cap C = \emptyset$, for all integers m. But since $\mu(C) > 0$ and no point of C returns to C, h cannot be μ-recurrent. $\qquad\square$

5.3 The Open Square

We begin our discussion of the case of the open square $S = (0,1)^2$ by recalling Brouwer's more famous fixed point theorem, that any continuous map of the closed square $[0,1]^2$ into itself has a fixed point. However, even for homeomorphisms of the closed square, there need be no *interior* fixed points, and so there are homeomorphisms of the open square with no fixed points. A simple example is the homeomorphism $(x_1, x_2) \mapsto (x_1^2, x_2)$. Note that this transformation pushes all points to the left and hence does not preserve area, or indeed any other finite measure. Since this transformation is obviously not recurrent, an easy way to verify the last observation is via the following well known but elementary result known as the Poincaré Recurrence Theorem, which says that if a transformation preserves a finite measure (i.e., is an *automorphism* of the measure space) then it is recurrent.

Theorem 5.4 (Poincaré Recurrence Theorem) *Any automorphism g of a finite measure space (X, Σ, μ) is μ-recurrent.*

Proof Suppose on the contrary that the (measurable) set of nonreturning points $B = A - \bigcup_{m \geq 1} g^{-m}(A)$ has positive measure, $\mu(B) > 0$. Observe that since $\mu\left(g^{m+k}(B) \cap g^k(B)\right) = \mu\left(g^m(B) \cap B\right) = 0$, the iterates $g^m(B)$, $m > 1$ must be disjoint (up to sets of measure zero). But this means that $\mu(X) \geq \mu\left(B \cup g(B) \cup g^2(B) \cup \cdots \cup g^{r-1}(B)\right) = r\,\mu(B)$ for any integer r, which implies that $\mu(X)$ is infinite, contrary to hypothesis.

$\square$

We note that for infinite measure spaces not all automorphisms are recurrent. For example simple translation by one unit to the right is a length preserving automorphism of the line which is not recurrent. However, we will find that if the invariant measure has no atoms (is zero for points) then all ergodic automorphisms are recurrent. But, these are matters for Part III.

Before finishing our discussion of fixed points, we cannot resist mentioning a paradoxical application of the Poincaré Recurrence Theorem to statistical mechanics. It is known (as Liouville's Theorem) that the dynamical system describing the motion of particles of an ideal gas preserves a finite measure which is the probability that the system is in a given state (set in phase space). Suppose all the particles in a room are pushed to the left side of the room by a moving partition of the room, and then this partition is removed and the gas continues its motion. Since the state A described by 'all the particles are in the left side of the room' has a very small but positive probability, Poincaré's Recurrence Theorem says that eventually (for the m with $\mu\left(g^m(A) \cap A\right) > 0$) there will be another time when all the particles are on the left side.

Returning to the fixed point problem, we can now prove the following application of Brouwer's Plane Translation Theorem, which was first observed in passing by Montgomery [86]. See also [32] and [43].

Theorem 5.5 *Let h be any orientation preserving, area preserving homeomorphism of the open square $S = (0,1)^2$. Then h has a fixed point.*

Proof Since h preserves a finite measure (area λ on the open square), Poincaré's Recurrence Theorem says that h is λ-recurrent. But the open square is homeomorphic to the plane, so that by Corollary 5.3 (of the Plane Translation Theorem) h must have a fixed point. $\square$

Bourgin [43] has asked whether this result can be extended to higher dimensions or orientation reversing homeomorphisms. We cannot answer this question now, but in the last section of Chapter 10 (see Theorem 10.8) we will show how the Homeomorphic Measures Theorem can be used to construct counterexamples.

5.4 The Torus

The second surface where the assumption of area preservation leads to
a fixed point is the torus T^2, which we will consider in the form of the
unit square with opposite sides identified. As in the previous chapter, we
will assume that distance is given by the maximum metric. An obvious
example of an area preserving homeomorphism of T^2 is the rotation by a
vector v, i.e., $x_i \mapsto x_i + v_i \pmod 1$. This rotation lifts to a translation by
v in the plane, and hence has mean rotation vector v. One way to avoid
this 'no fixed point' example, which turns out to avoid all such examples,
is to require mean rotation zero. As techniques of this section work
equally well on the annulus A obtained from the square by identifying
opposite sides in the first coordinate only, we will include the case that
the underlying space is the annulus as an alternative hypothesis. This
will be the relevant case for the next section, which considers fixed points
of homeomorphisms of the annulus.

Theorem 5.6 (Franks) *An area preserving homeomorphism f of the
torus or the annulus which is homotopic to the identity and has mean
rotation zero has a fixed point.*

Before proving this result it is worth tracing its history. It was conjectured by Arnold (see [34] and [35, Appendix 9]), and later proved
(along with more general results) by Conley and Zehnder [53], that any
diffeomorphism of the torus which has mean rotation zero must have at
least three fixed points. Franks [62] gave a direct proof that weaken-
ing the smoothness hypothesis from diffeomorphism to homeomorphism
would still keep a fixed point; and then Flucher [60] showed that with
Franks's hypotheses there are at least two fixed points. The existence
of a third fixed point under these assumptions has recently been shown
by P. Le Calvez [81].

The shortest proof of Franks's Theorem runs along the following lines.
Suppose f is a torus homeomorphism satisfying Franks's hypotheses
which has no fixed point. Then all sufficiently close homeomorphisms h,
as well as their lifts $\hat{h}$ to the plane, will also have no fixed points. Hence
if C is a sufficiently small closed disk, the condition (5.1) ensures that the
$\hat{h}$ iterates of C are disjoint. So an $\hat{h}$-orbit can enter C at most once and
consequently cannot be dense in C (or in the plane). So $\hat{h}$ is not transitive. It follows that a neighborhood of f in $\mathcal{M}^0[T^2, \lambda]$ has nontransitive
lifts to the plane, contradicting our earlier result, Theorem 4.2.

We prefer however to give the following self-contained proof of Franks's Theorem based on the discrete methods of Chapter 3, which does not require the Baire Category Theorem.

Proof We use Franks's idea of uniformly approximating f by a homeomorphism h, whose lift $\hat{h}$ to the plane has a periodic point. Our combinatorial proof of this fact is given in Theorem 3.7. Hence by the Plane Translation Theorem (Theorem 5.1) $\hat{h}$, and therefore h as well, must have a fixed point. Since f can be uniformly approximated by homeomorphisms with fixed points, it follows from the fact that the torus is compact that f itself must have a fixed point.   □

5.5 The Annulus

In the same year, 1912, that Brouwer published his investigations on plane homeomorphisms, Poincaré published the paper containing his celebrated 'Last Geometric Theorem'. He conjectured (and gave proofs for certain cases) that an orientation preserving, area preserving homeomorphism of the annulus which rotates the bounding circles in opposite directions must have a fixed point. The condition on the bounding circles is clearly needed to exclude rotations, which have no fixed points. Birkhoff's work on this problem [41] proved the general case, and established the existence of a second fixed point. Recently, Franks [61] has weakened the assumption of area preservation. The interested reader may also wish to consult the recent book by Pollicott and Yuri, *Dynamical systems and ergodic theory* for further details [95] (see also [82] for another approach to these results).

We will prove the following version of the Last Geometric Theorem, first for the special case where the bounding circles are each rotated by some fixed angle and then (referring to results from Chapter 9 on homeomorphic measures) for the more general stated condition on the circles.

Theorem 5.7 *Let f be an area preserving homeomorphism of the annulus $A = S^1 \times [0,1]$, homotopic to the identity, with a lift $\hat{f}$ to $R^1 \times [0,1]$ satisfying one of the following conditions for all $x \in R^1$:*

(i) $\hat{f}(x,0) = (x-a,0)$, $\hat{f}(x,1) = (x+b,1)$, $a,b > 0$,

or the weaker condition

(ii) *The first coordinate* $\hat{f}_1(x,0) < x$ *and* $\hat{f}_1(x,1) > x$.

Then the homeomorphisms $\hat{f}$ *and* f *have fixed points.*

Proof We begin with a self-contained proof for the simpler case (i). Note that if $v_t(\hat{f}) = 0$, where v_t is the mean translation vector defined in formula (3.5), the result is already included in Franks's Theorem (Theorem 5.6) above. So without loss of generality we assume that $v_t(\hat{f}) < 0$. In this case extend $\hat{f}$ to a homeomorphism $\hat{g}$ of the strip $R^1 \times [0, 1 - v_t(\hat{f})/b]$ by defining $\hat{g}(x,y) = (x + b, y)$ for $y > 1$. Since $\hat{g}$ is a rotationless orientation preserving area preserving homeomorphism of a strip $R^1 \times J$ (J an interval) it follows from the proof of Franks's Theorem above that it has a fixed point. Since the fixed point must lie below the line at height 1, it projects onto a fixed point of f.

In the more general case (ii) we first extend f to an area preserving homeomorphism h of the larger annulus $S^1 \times [-1, 2]$, such that h satisfies condition (i) for the circles at height $-1, 2$. The existence of such an extension follows from the Homeomorphic Measures Theorem. While h has a fixed point by part (i), we cannot assert it lies in the original annulus, so a slight modification is needed. Let $\hat{h}$ be the lift of h to $R^1 \times [-1, 2]$ which extends $\hat{f}$. Let $\hat{s}$ be a shear homeomorphism of $R^1 \times [-1, 2]$ (a translation on each horizontal line) which is the identity on $R^1 \times [0, 1]$ and moves every line above 1 sufficiently far to the right and every line below 0 sufficiently far to the left so that the composition $\hat{s}\hat{h}$ moves every point above 1 to the right and every point below 0 to the left. Thus if $\hat{s}\hat{h}$ has a fixed point, it projects onto a fixed point of f in the original annulus $A = S^1 \times [0, 1]$. But the homeomorphism $\hat{s}\hat{h}$ satisfies the hypotheses of case (i), and the existence of a fixed point in this case has been shown above. $\qquad\qquad\square$

6

Measure Preserving Lusin Theorem

6.1 Introduction

A central idea of real variable theory, 'Littlewood's Second Principle', is that every measurable function is nearly continuous. Two forms of this principle are contained in the following well known result, the stronger second part of which is known as 'Lusin's Theorem'.

Theorem 6.1 *Let $g : R \to R$ be a measurable real valued function with $|g(x) - x| < \epsilon$ on the interval $[a, b]$. Then for any $\delta > 0$ there is a continuous function $h : R \to R$ with $|h(x) - x| < \epsilon$ on $[a, b]$ satisfying*

(i) $\lambda \{x : |g(x) - h(x)| \geq \delta\} < \delta$, *and even*
(ii) $\lambda \{x : g(x) \neq h(x)\} < \delta$.

In this chapter we will prove an analogous result which relates measurable and continuous ergodic theory. That is, we show that a volume preserving bimeasurable bijection of the cube I^n is nearly a volume preserving homeomorphism. The notion of 'nearly' is made precise in the following result obtained by Alpern [8].

Theorem 6.2 (Measure Preserving Lusin Theorem) *Let g be a bimeasurable volume preserving bijection (i.e., automorphism) of the cube I^n, $n \geq 2$, with $\|g\| \equiv \operatorname{ess\,sup} |g(x) - x| < \epsilon$. Then given any $\delta > 0$, there is a volume preserving homeomorphism h of I^n, with $\|h\| < \epsilon$ and equal to the identity on the boundary of I^n, satisfying*

(i) $\lambda \{x : |g(x) - h(x)| \geq \delta\} < \delta$
(ii) $\lambda \{x : g(x) \neq h(x)\} < \delta$.

We shall only prove the weaker part (i) in this book, as it has many important consequences in the ergodic theory of measure preserving homeomorphisms, and the stronger part (ii) is more difficult yet has few additional known consequences. We should also note that while in the real variables result (Theorem 6.1) the preservation of the uniform ϵ bound is a trivial afterthought, in our context it is crucial to the applications and not easy to prove.

Our Theorem 6.2 can be viewed more abstractly as saying that the space $\mathcal{M}[I^n, \lambda]$ of volume preserving homeomorphisms of I^n is a dense subset of the space $\mathcal{G}[I^n, \lambda]$ with respect to the *weak topology* given by the *weak metric* $\rho(f, g) = \inf\{\delta : \lambda\{x : |f(x) - g(x)| \geq \delta\} < \delta\}$. It can be shown that this topology is the same as that given without reference to the topology of I^n, by saying that a sequence g_i converges to a limit g if and only if $\lambda(g_i(A) \triangle g(A)) \to 0$ for every measurable set A [8]. Recall that both the uniform topology on $\mathcal{M}[I^n, \lambda]$ (given by the distance $\|h - h'\|$) and the weak topology on $\mathcal{G}[I^n, \lambda]$ (see [72]) are topologically complete. We shall also consider the uniform topology on $\mathcal{G}[I^n, \lambda]$ – it is complete with this topology. The space $\mathcal{G}[I^n, \lambda]$ with the weak topology was extensively studied by Halmos in two 1944 papers [69] and [70]. In the second paper he applied Baire category arguments to show that the set of weak mixing transformations is *generic* in the weak topology. The approximation theorem of this chapter will enable us to establish similar and more general results for the space $\mathcal{M}[I^n, \lambda]$ with the uniform topology, using the following corollary of Theorem 6.2 (which we will prove later).

Corollary 6.3 *Let $\mathcal{V}$ be a G_δ subset of $\mathcal{G}[I^n, \lambda]$ in the weak topology. If the uniform topology closure of $\mathcal{V}$ contains $\mathcal{M}[I^n, \lambda]$, then $\mathcal{V} \cap \mathcal{M}[I^n, \lambda]$ is a dense G_δ subset of $\mathcal{M}[I^n, \lambda]$, in the uniform topology. In particular, $\mathcal{V} \cap \mathcal{M}[I^n, \lambda]$ is nonempty. This result remains true if $\mathcal{M}[I^n, \lambda]$ is replaced by $\mathcal{M}[I^n, \partial I^n]$, the subset of $\mathcal{M}[I^n, \lambda]$ consisting of homeomorphisms equal to the identity on ∂I^n, the boundary of I^n.*

Proof We leave the proof of the final sentence regarding $\mathcal{M}[I^n, \partial I^n]$ to the end. For the results on $\mathcal{M}[I^n, \lambda]$ we first prove the result with 'G_δ' replaced by 'open' in both instances. So assume that $\mathcal{V}$ is open in the weak topology. Since the uniform topology is finer than the weak topology, $\mathcal{V} \cap \mathcal{M}[I^n, \lambda]$ is open in the uniform topology on $\mathcal{M}[I^n, \lambda]$. Theorem 6.2 and the hypotheses imply that $\mathcal{V}$ is dense in $\mathcal{G}[I^n, \lambda]$ in the weak topology. To show that $\mathcal{V} \cap \mathcal{M}[I^n, \lambda]$ is dense in $\mathcal{M}[I^n, \lambda]$ in the uniform topology, it suffices to show that $\mathcal{V} \cap \mathcal{M}[I^n, \lambda] \cap \mathcal{B} \neq \emptyset$ for every uniform metric ball $\mathcal{B} = \mathcal{B}(f; \epsilon) = \{g \in \mathcal{G}[I^n, \lambda] : \|gf^{-1}\| < \epsilon\}$ in $\mathcal{G}[I^n, \lambda]$

centered at a homeomorphism f in $\mathcal{M}[I^n, \lambda]$. By hypothesis there is some g_0 in $\mathcal{V} \cap \mathcal{B}$. Therefore $\|g_0 f^{-1}\| < \epsilon$, and also $g_0 f^{-1}$ belongs to the weak topology open set $\mathcal{V} f^{-1} = \{c f^{-1}, c \in \mathcal{V}\}$. By Theorem 6.2 applied to the transformation $g_0 f^{-1}$, there is a homeomorphism $h \in \mathcal{M}[I^n, \lambda]$ which equals the identity on the boundary of the cube, belongs to the weak open neighborhood $\mathcal{V} f^{-1}$ and satisfies $\|h\| < \epsilon$. Consequently hf belongs to the set $\mathcal{V} \cap \mathcal{M}[I^n, \lambda] \cap \mathcal{B}$ which is thus nonempty, as required.

We now consider the 'G_δ' case, that is, we assume that $\mathcal{V}$ is a weak topology G_δ set. The facts that $\mathcal{V}$ is dense in $\mathcal{G}[I^n, \lambda]$ in the weak topology and that $\mathcal{V} \cap \mathcal{M}[I^n, \lambda]$ is a G_δ subset of $\mathcal{M}[I^n, \lambda]$ in the uniform topology follow as in the 'open' case. To prove the rest, represent $\mathcal{V}$ as the countable intersection of weak open sets $\mathcal{V}_i$. Observe that each open set $\mathcal{V}_i$ satisfies the assumptions of the 'open' case of this corollary already proved, so we may conclude that each set $\mathcal{V}_i \cap \mathcal{M}[I^n, \lambda]$ is dense and open in the uniform topology on $\mathcal{M}[I^n, \lambda]$. Hence the Baire Category Theorem asserts that the intersection

$$\mathcal{V} \cap \mathcal{M}[I^n, \lambda] = \bigcap_{i=1}^{\infty} (\mathcal{V}_i \cap \mathcal{M}[I^n, \lambda])$$

of these dense open sets is a dense G_δ set, as required.

The proof for the statement where $\mathcal{M}[I^n, \lambda]$ is replaced by $\mathcal{M}[I^n, \partial I^n]$ is essentially the same; only the first paragraph must be changed slightly. When we choose the target homeomorphism f, we may additionally assert that $f \in \mathcal{M}[I^n, \partial I^n]$. Since the proof in the first paragraph asserts that h is the identity on ∂I^n, the boundary of I^n, it follows that hf is also the identity on ∂I^n, so that hf belongs to the smaller set $\mathcal{V} \cap \mathcal{M}[I^n, \partial I^n] \cap \mathcal{B}$. $\square$

The importance of this corollary is that it gives us a way to establish the existence (and typicality) of homeomorphisms with specified ergodic theoretic properties described by the subset $\mathcal{V}$ of $\mathcal{G}[I^n, \lambda]$. For example, we may take $\mathcal{V}$ to be the set of all ergodic or perhaps weak mixing transformations – sets known to be defined by a countable number of weakly open conditions (and hence G_δ). Corollary 6.3 enables us to do our approximations (in establishing density) by *discontinuous* automorphisms with properties such as ergodicity.

What remains to be shown then is the second condition, that an arbitrary volume preserving homeomorphism can be uniformly approximated by a transformation in $\mathcal{G}[I^n, \lambda]$ (not necessarily continuous) with the required property (membership in $\mathcal{V}$). In the next chapter we will show how an arbitrary volume preserving homeomorphism of I^n can be uniformly approximated by a (discontinuous) ergodic transformation of I^n. This corollary then will imply the existence and general nature of ergodic volume preserving homeomorphisms of I^n. We shall later use this corollary to establish the existence of volume preserving homeomorphisms with much more general measure theoretic properties.

6.2 Approximation Techniques

In this section we outline the techniques of approximation that will be needed in the next section to prove Theorem 6.2(i). It may be useful at this point to recall the method of proof for the real variable result of Theorem 6.1(i). There, the measurable function is first approximated by a step function, which is in turn approximated by a continuous function. Here, the role of intermediate is played by dyadic permutations. The rough outline of the proof, given a transformation $g \in \mathcal{G}[I^n, \lambda]$ with $\|g\| < \epsilon$, is as follows:

step 1 Approximate g by a dyadic permutation R with $\rho(g, R)$ small. The technique for this approximation is given in Lemma 6.4 below. Clearly most dyadic cubes will also move less than ϵ, but a small fraction (the 'bad' cubes) may move further, so that $\|R\|$ may be large.

step 2 Approximate R by another dyadic permutation P with $\rho(R, P)$ small and $\|P\| < \epsilon$. This is accomplished by setting P equal to the identity on the 'bad' cubes of R, and modifying R on the remaining dyadic cubes so that their images avoid the small set of bad cubes. A 1-dimensional version of this process is given in Lemma 6.5, where $\hat{Q}$ will be R^{-1} on the set F of images of the bad cubes and $P = QR$ is the required dyadic permutation. The extension to higher dimensions is achieved by numbering the dyadic cubes so that close indices imply close cubes.

step 3 Approximate the dyadic permutation P by a volume preserving homeomorphism $h \in \mathcal{M}[I^n, \partial I^n]$ with $\|h\| < \epsilon$ and $\rho(P, h)$ small. The technique for this step is given in Lemma 6.6 below, which relies on constructions from Chapter 2.

We now present the three lemmas corresponding to the three steps above in order to prove Theorem 6.2, which will itself be proved in the next section. The following lemma was first proved by Halmos in [69,

p. 6] (see also [72]) though in a different way. In his paper [69, footnote 14], Halmos notes that the result below

... has the rank of a 'folk theorem'. .... It has a satisfying intuitive content: it says that, in the limit, every measure preserving transformation is obtained by cutting up the space (with an ordinary pair of Euclidean scissors) into a finite number of pieces and then merely permuting the pieces. The first precise formulation of this result (not the one below) I heard from John von Neumann in November 1940. ... The first published version (different from both von Neumann's and mine) is due to Oxtoby and Ulam [88, p. 919].

The theorem of Oxtoby and Ulam referred to by Halmos above is [88, Theorem 12] and generalized in Theorem 4.9. The proof given here is based on [8].

Lemma 6.4 *The dyadic permutations are dense in* $\mathcal{G}[I^n, \lambda]$, *in the weak topology. That is, given* $g \in \mathcal{G}[I^n, \lambda]$ *and positive numbers* δ *and* γ, *there is a dyadic permutation* $R \in \mathcal{G}[I^n, \lambda]$ *with*

$$\lambda \{x : |g(x) - R(x)| \geq \delta\} < \gamma.$$

Proof Choose a dyadic decomposition σ_i, $i = 1, \ldots, N$, of I^n with diameter less than δ. Let C_i, $i = 1, \ldots, N$, be disjoint compact sets with $C_i \subset g^{-1}(\sigma_i)$ and $\lambda \left(\bigcup_{i=1}^N C_i \right) > 1 - \gamma/2$. Let $d > 0$ denote the minimum distance between points in distinct sets C_i and let τ_j, $j = 1, \ldots, M$, be any dyadic decomposition with cubes of diameter less than d and volume less than $1/N - \max_i \lambda(C_i)$ that refines the σ_i. Thus no dyadic cube τ_j can intersect more than one of the sets C_i.

We now assign to each set C_i a collection of the cubes τ_j which intersect it and have total measure between $\lambda(C_i)$ and $1/N$. Let S_i be the union of these cubes. Now define R as a permutation of the τ_j by first defining it on cubes in the set S_i so that if $\tau_j \subset S_i$ then $R(\tau_j) \subset \sigma_i$, and then extending it arbitrarily to the remaining cubes.

Since $S_i \cap C_{i'} = \emptyset$ for $i \neq i'$ we have

$$\bigcup_{i=1}^N (C_i \cap S_i) = \left(\bigcup_{i=1}^N C_i \right) \cap \left(\bigcup_{i=1}^N S_i \right).$$

Call this common set S and observe from the right side of the equation that $\lambda(S) > 1 - \gamma$. Furthermore it follows from the left side that if $x \in S$ then for some i, $x \in C_i \cap S_i$ so $g(x) \in \sigma_i$ and $R(x) \in \sigma_i$. Therefore for $x \in S$, $|g(x) - R(x)| < \delta$, as required.   □

Lemma 6.5 *Let $F \subset \{1, \ldots, N\}$ and let $\hat{Q} : F \to \{1, \ldots, N\}$ be injective. Then there is a permutation $Q : \{1, \ldots, N\} \to \{1, \ldots, N\}$ which extends $\hat{Q}$ and satisfies $|j - Q(j)| \leq \#(F)$ for $j \notin F$, where $\#(F)$ denotes the cardinality of the set F.*

Proof Define Q as $\hat{Q}$ on F, and $\check{Q}$ on the complement $\tilde{F}$, where $\check{Q}$ on the complement of F is the unique order preserving bijection onto the complement of QF (i.e., $\check{Q} : \tilde{F} \to \widetilde{QF}$). That is, if $j \in \tilde{F}$ and $\check{Q}j = j'$, then

$$q = \#\{k : k \in \tilde{F} \text{ and } k \leq j\} = \#\{l : l \in \widetilde{\hat{Q}F} \text{ and } l \leq j'\}.$$

It follows that $j - \#(F) \leq q \leq j$ and $j' - \#(\hat{Q}F) \leq q \leq j'$. But since $\#F = \#(\hat{Q}F)$ we have that $j - \#(F) \leq j'$ and $j' - \#(F) \leq j$, or $|j - j'| \leq \#(F)$, which is the required estimate. $\qquad\square$

The next lemma says that the strong topology closure of $\mathcal{M}[I^n, \partial I^n]$ (the volume preserving homeomorphisms of I^n which fix the boundary ∂I^n) in $\mathcal{G}[I^n, \lambda]$ contains the set of all dyadic permutations. The strong topology is given by the metric which defines the distance between two automorphisms f and g by $\lambda\{x : f(x) \neq g(x)\}$. It is obviously finer than the weak topology.

The following result is a special case of Theorem 6.2(ii) with g restricted to dyadic permutations, and constitutes what we earlier called **step 3**.

Lemma 6.6 *Given a dyadic permutation P of I^n with $\|P\| < \epsilon$, and a positive number γ, there is a volume preserving homeomorphism $h \in \mathcal{M}[I^n, \lambda]$, $\|h\| < \epsilon$, which equals the identity on the boundary and satisfies*

$$\lambda\{x : P(x) \neq h(x)\} < \gamma.$$

Proof View P as a permutation of dyadic cubes σ_i, $i = 1, \ldots, N$, with diameter less than $(\epsilon - \|P\|)/3$. For $0 < \beta < 1$, let σ_i^β denote the cubes concentric to σ_i, with parallel faces, and with a fraction β of the measure of σ_i. We now apply Theorem 2.4 to the N centers of the σ_i, and to their images under P, which are by assumption within a distance ϵ. Since $\|P\| < \epsilon - \frac{2}{3}(\epsilon - \|P\|)$ we obtain in this way a homeomorphism $f \in \mathcal{M}[I^n, \partial I^n]$ with $\|f\| < \epsilon - \frac{2}{3}(\epsilon - \|P\|)$; furthermore there is some positive number α, with $0 < \alpha < 1$, such that the homeomorphism f agrees with P pointwise on every cube σ_i^α and thus $\lambda\{x : P(x) \neq f(x)\} \leq 1 - \alpha$. Unfortunately this α may be too small. However, if we could find a homeomorphism h which agreed with P on the cubes σ_i^β, for some β large enough so that $\beta > 1 - \gamma$, the lemma would be proved. Take such a number β (i.e., $\beta > 1 - \gamma$). To obtain h we will define a homeomorphism (not volume preserving) $T : I^n \to I^n$, which leaves each cube σ_i invariant, maps each cube σ_i^α radially onto the concentric cube σ_i^β, and has constant Jacobian β/α on the f-invariant set $A = \bigcup_{i=1}^N \sigma_i^\alpha$ and constant Jacobian $(1 - \beta)/(1 - \alpha)$ on $\tilde{A}$, the complement of A. The composition $h = TfT^{-1}$ will then have the required properties.

The construction of T reduces to the construction of homeomorphisms $T_i : \sigma_i \to \sigma_i$ which fix the boundary of σ_i, and furthermore $T_i : \sigma_i^\alpha \to \sigma_i^\beta$ with constant scaling β/α on σ_i^α and constant Jacobian $(1 - \beta)/(1 - \alpha)$ on $\sigma_i - \sigma_i^\alpha$. The T_i together define T by $T(x) = T_i(x)$ for $x \in \sigma_i$. To this end we define $T_i : \sigma_i \to \sigma_i$ to be the homeomorphism which maps the boundary of σ_i^t linearly (radially) onto the boundary of the concentric cube $\sigma_i^{r(t)}$, where $r : [0,1] \to [0,1]$ is the piecewise linear function determined by $r(0) = 0$, $r(\alpha) = \beta$ and $r(1) = 1$. Note that since T leaves each cube σ_i invariant, it follows that $\|T\| \leq \operatorname{diam}(\sigma_i) < (\epsilon - \|P\|)/3$.

We now check that $h = TfT^{-1}$ has the required properties. It follows immediately from the definitions of f and T that $h = P$ on each cube σ_i^β, and hence that $\lambda\{x : P(x) \neq h(x)\} \leq 1 - \beta < \gamma$. To establish that h preserves volume, we need only check that $\lambda(h(S)) = \lambda(S)$ for each subset S of the complement of $\bigcup_i \sigma_i^\beta$. For such a set S we see that

$$
\begin{aligned}
\lambda(h(S)) &= \lambda\left(TfT^{-1}(S)\right) \\
&= \frac{1-\beta}{1-\alpha}\lambda\left(fT^{-1}(S)\right) \\
&= \frac{1-\beta}{1-\alpha}\lambda\left(T^{-1}(S)\right) \\
&= \frac{1-\alpha}{1-\beta}\frac{1-\beta}{1-\alpha}\lambda(S) \\
&= \lambda(S).
\end{aligned}
$$

Finally, we calculate that

$$\|h\| \leq \|f\| + 2\|T\|$$
$$< \epsilon - \frac{2}{3}(\epsilon - \|P\|) + \frac{2}{3}(\epsilon - \|P\|) = \epsilon.$$

$\square$

Note that we can give the following 'physical' model of the action of $h = TfT^{-1}$ in the previous lemma. Consider the cube I^n as consisting of N ice cubes in positions σ_i^β, $i = 1, \ldots, N$, and filled up with water elsewhere. The transformation T^{-1} in the proof changes the temperature so that the ice cubes σ_i^β shrink uniformly in volume to σ_i^α and the water expands uniformly in volume so that I^n is completely filled with smaller ice cubes and more water. Furthermore the temperature change T is effected so that each σ_i is invariant. The shrunken ice cubes σ_i^α are moved by the measure preserving homeomorphism f rigidly to their new positions $P(\sigma_i^\alpha)$. Next we bring the temperature back to normal (i.e., apply T to I^n), so that the small ice cubes σ_i^α grow back to their original size σ_i^β. The composition of these maps (the temperature changes T and T^{-1} and the rigid motion f) is volume preserving (since the initial and final temperatures are the same) and equals P on all of the σ_i^β.

6.3 Proof of Theorem 6.2(i)

In this section we combine the various approximation techniques of the previous section to complete the proof of the first part of Theorem 6.2. The second part of Theorem 6.2, the 'Lusin Theorem for Measure Preserving Homeomorphisms', uses similar ideas but also more complicated topological ideas that have little to do with the assumption of volume preservation, and hence lies outside the scope of this book. In addition, we shall have no need for the stronger result in the rest of the book. The full proof of that result can be found in [8] which uses many results from [19]. Earlier proofs of this result without the norm preservation clause (which is essential for our applications) are due to Oxtoby [90] and White [107].

Proof of Theorem 6.2(i) To avoid confusion between the first δ (which represents distance) and the second δ (which represents volume), we will use two constants, constructing an $h \in \mathcal{M}[I^n, \partial I^n]$ which satisfies $\lambda\{x : |g(x) - h(x)| \geq \delta\} < \gamma$, for arbitrary positive δ and γ. We begin the construction of h by first fixing a dyadic decomposition $\{\tau_i\}_{i=1}^L$ with diameter less than $\alpha/2$ where

$$\alpha < \max\left[\left(\epsilon - \|g\|\right)/2, \delta/2\right].$$

For later purposes we fix a numbering of the τ_i so that cubes with consecutive indices share a common face. Next choose $\beta > 0$ with $\beta < \min\left[1/L, \gamma\right]$. By Lemma 6.4, there is a dyadic permutation R which weakly approximates the given transformation $g \in \mathcal{G}[I^n, \lambda]$, in the sense that

$$\lambda\left\{x : |g(x) - R(x)| \geq \alpha\right\} < \beta. \tag{6.1}$$

We may view R as a dyadic permutation of a decomposition $\{\sigma_j\}_{j=1}^N$, where $N = KL$ and thus σ_j refines the τ_i. Number the cubes σ_j so that $\sigma_j \subset \tau_i$ if and only if $(i-1)K < j \leq iK$. The numberings of the τ_i and the σ_j together ensure that if $|j_1 - j_2| \leq K$ then σ_{j_1} and σ_{j_2} are in adjacent elements of the τ_i and therefore

$$\operatorname{diam}\left(\sigma_{j_1} \cup \sigma_{j_2}\right) < \alpha \quad \text{if } |j_1 - j_2| \leq K. \tag{6.2}$$

We now wish to weakly approximate the dyadic permutation R by another dyadic approximation P, with $\|P\| < \epsilon$, a process referred to as **step 2** in the last section. This involves defining P to be the identity on the 'bad cubes' σ_j belonging to the set $D = \{x : |x - R(x)| \geq \|g\| + \alpha\}$, and as close as possible to R on the remaining cubes. Note that the set D is the union of dyadic cubes (since on each cube R is a translation and so $R(x) - x$ is constant there). First observe that there are relatively few bad cubes, in the sense that $\lambda(D) < \beta < 1/L$ by (6.1) and the choice of β. Since each cube σ_j has volume $1/N = 1/(KL)$ this estimate on the volume of D implies that D consists of fewer than $K = N/L$ of the σ_j. Let $\hat{Q} : R(D) \to I^n$ be the restriction of R^{-1} to $R(D)$, so that $\hat{Q}R$ will be the identity on the bad cubes D. Since $\hat{Q}$ is a permutation of the N cubes σ_j we may view it as a map of the cube indices, i.e., $\hat{Q} : F \to \{1, \ldots, N\}$ where $F = \{j : R^{-1}(\sigma_j) \subset D\}$.

Now apply Lemma 6.5 to the injection $\hat{Q} : F \to \{1, \ldots, N\}$, to obtain an extension (permutation) $Q : \{1, \ldots, N\} \to \{1, \ldots, N\}$ with $|j - Q(j)| \leq \#(F) \leq K$ for all $j \notin F$. If we now view Q as a dyadic permutation of the σ_j, we claim that the dyadic permutation $P = QR$ satisfies

$$\|P\| < \epsilon, \text{ and} \tag{6.3}$$

$$\lambda\left\{x : |g(x) - P(x)| \geq \delta\right\} < \gamma. \tag{6.4}$$

If we can establish these two properties of P, then we can apply Lemma 6.6 to find a homeomorphism $h \in \mathcal{M}[I^n, \partial I^n]$ with small norm, $\|h\| < \epsilon$, and close to P, $\lambda \{x : P(x) \neq h(x)\} < \gamma_0$ where γ_0 is the difference between the right and left sides of the inequality (6.4). Hence we will have $\lambda \{x : |g(x) - h(x)| \geq \delta\} < \gamma$, as required.

It remains only to establish the two properties (6.3) and (6.4). To check the first of these recall that by definition P equals the identity on the 'bad cubes' (those in D). On the other hand suppose that $y \notin D$, and that $R(y) \in \sigma_{j_1}$. Then $QR(y) \in \sigma_{j_2}$ for some j_2 with $|j_1 - j_2| \leq K$ and hence $|R(y) - QR(y)| < \alpha$ by (6.2). It follows that

$$\begin{aligned} |y - QR(y)| &\leq |y - R(y)| + |R(y) - QR(y)| \\ &\leq (\|g\| + \alpha) + \alpha \leq \epsilon, \end{aligned}$$

by the definition of D and the initial choice of α. This proves (6.3). To establish (6.4), first consider any point y belonging to the set $S = \{x : |g(x) - R(x)| < \alpha\} \subset \tilde{D}$ with $\lambda(S) \geq 1 - \beta > 1 - \gamma$ (by (6.1)), where R is a good approximation to g. Observe that

$$\begin{aligned} |g(y) - P(y)| &\leq |g(y) - R(y)| + |R(y) - P(y)| \\ &\leq \alpha + \alpha < \delta. \end{aligned}$$

It follows that $\lambda \{x : |g(x) - P(x)| \geq \delta\} < \gamma$, establishing the last required estimate (6.4). $\qquad\square$

The approach to Lusin's Theorem for measure preserving homeomorphisms taken by Oxtoby [90] and White [107] is quite different. Given a volume preserving automorphism g of the cube, they use a general (i.e., one without a measure preserving assumption) Lusin Theorem to assert that for a set A of *special topological type* and measure arbitrarily close to that of the cube, the restriction $\hat{g}$ of g to A is continuous. The *special topological type* (a Cantor set, in [90]; a sectionally 0-dimensional set in [107]) is chosen with the property that a homeomorphism $\hat{g}$ with domain of that type may be extended to a homeomorphism h of the cube. The fact that the extension may also be done in a volume preserving manner (with $h \in \mathcal{M}[I^n, \lambda]$) is established using the Homeomorphic Measures Theorem (our Theorem 9.1). However, it is *not* the case that a homeomorphism between (say) Cantor sets in the cube which moves no point a distance of more than ϵ may be extended to a homeomorphism of the whole cube with that property. Thus the norm preserving aspect of our Theorem 6.2 cannot be established via the methods of Oxtoby or White.

7

Ergodic Homeomorphisms

7.1 Introduction

The Lusin Theorem (or rather its consequence Corollary 6.3) in the previous chapter provides us with a method of constructing volume preserving homeomorphisms with desired measure theoretic properties. This method reduces the problem to approximating a volume preserving homeomorphism uniformly by a volume preserving automorphism (not necessarily continuous) with the desired measure theoretic property. In the next chapter we will give a very general application of this method, but here we use it simply to demonstrate the existence (and typicality) of ergodic homeomorphisms of the cube. (We recall that an automorphism of a finite measure space is said to be ergodic if its only invariant sets are of measure zero or full measure.) Again, this is an optional chapter, in that a stronger result (Theorem 8.2) will be proved independently in the next chapter.

However, the proof we present here, that ergodicity is typical among volume preserving homeomorphisms of the cube, is a very clear illustration of the method of approximation by discontinuous automorphisms. Given Corollary 6.3 of the previous chapter, we are required only to approximate an arbitrary homeomorphism in $\mathcal{M}[I^n, \lambda]$ by an ergodic (generally discontinuous) *automorphism* in $\mathcal{G}[I^n, \lambda]$, in the uniform topology.

Theorem 7.1 *The ergodic homeomorphisms form a dense G_δ subset of the volume preserving homeomorphisms of I^n, in the uniform topology.*

Proof Let $\mathcal{G}[I^n, \lambda]$ denote the space of all volume (λ) preserving bimeasurable bijections of the unit cube, endowed with the weak topology.

Since the set $\mathcal{E} \subset \mathcal{G}[I^n, \lambda]$ consisting of ergodic automorphisms is a G_δ set (see [72]), Corollary 6.3 reduces the problem to showing the following: *given any $h \in \mathcal{M}[I^n, \lambda]$ and any $\epsilon > 0$, there is an ergodic automorphism $f \in \mathcal{G}[I^n, \lambda]$ with $|h(x) - f(x)| < \epsilon$ for λ-a.e. x in I^n.*

By Theorem 3.3, there is a cyclic dyadic permutation $P \in \mathcal{G}[I^n, \lambda]$ of some dyadic decomposition σ_i, $i = 1, \ldots, N$, such that $\|h - P\| +$ diameter $(\sigma_i) < \epsilon$. Number the cubes σ_i so that $P(\sigma_i) = \sigma_{i+1}$ for $i < N$, and $P(\sigma_N) = \sigma_1$. Let $\hat{f}$ be any ergodic volume preserving bijection of the cube σ_1, extended to an element of $\mathcal{G}[I^n, \lambda]$ by setting it equal to the identity off σ_1. (Note that we are *not* assuming that $\hat{f}$ is a homeomorphism.) Define $f = \hat{f}P$. The automorphism f may be viewed by stacking the cubes (think of squares) above one another, with base σ_1 and top σ_N. A point not in the top cube moves linearly up to the next cube under f, unless it is in the top cube – in which case it moves linearly to the bottom cube and then moves within the bottom cube according to the ergodic automorphism $\hat{f}$. For such constructions (and more general ones called skyscraper constructions – see Theorem A1.2 in Appendix 1) the resulting automorphism is ergodic if the base automorphism is ergodic. To see this, suppose that there is an f-invariant set S (that is, $f(S) = S$), with $0 < \lambda(S) < 1$. Then the sets $S_i = S \cap \sigma_i$ also have intermediate volume. It follows from the definitions of f and the S_i that $f^N(S_i) = S_i$. But $f^N(S_1) = \hat{f}(S_1)$ so that S_1 is a nontrivial invariant set of $\hat{f}$, contradicting the assumed ergodicity of $\hat{f}$. Hence our assumption that f is not ergodic (the existence of the set S) is false. Thus f is an ergodic automorphism with $\|P - f\| <$ diameter(σ_i) and hence $\|h - f\| < \epsilon$, as required.

$\square$

The above theorem, originally a conjecture of Birkhoff, is due to Oxtoby and Ulam [88] and is the main result of their theory. It is clearly stronger than Oxtoby's earlier corresponding result for transitivity [89], since ergodic homeomorphisms are necessarily transitive. The proof given here (not previously published) is very easy precisely because the Lusin Theory of Chapter 6 enables us to go outside of $\mathcal{M}[I^n, \lambda]$ (i.e., to use discontinuous automorphisms) to construct our ergodic approximations. The use of the Lusin Theory here (our Corollary 6.3) is essentially the same as in the next chapter, where we approximate by automorphisms with more restrictive measure theoretic properties, of which ergodicity and weak mixing are but special examples.

7.2 A Classical Proof of Generic Ergodicity

The proof of Theorem 7.1 (Generic Ergodicity) given in the previous section of this chapter relies heavily on the embedding of $\mathcal{M}[I^n, \lambda]$ in $\mathcal{G}[I^n, \lambda]$ and the consequent ability to use discontinuous automorphisms (elements of $\mathcal{G}[I^n, \lambda]$) in the approximation process. In particular, the above proof used dyadic permutations, which are not continuous. It also used the fact that the ergodic automorphisms form a G_δ subset of $\mathcal{G}[I^n, \lambda]$ with respect to the weak topology, and the existence of an ergodic automorphism. We now present, for historical comparisons, a version of the original proof of Theorem 7.1 given by Oxtoby and Ulam [88] as modified in [6]. Note that all the constructions take place within the space $\mathcal{M}[I^n, \lambda]$, and that all automorphisms are homeomorphisms. Note that nowhere is the existence of any ergodic automorphism assumed, nor any facts about $\mathcal{G}[I^n, \lambda]$ or the weak topology required.

Recall that in our modern proof of generic ergodicity (Theorem 7.1) given above, the dyadic permutations played a prominent role in the approximation argument. Since the dyadic permutations are not continuous, the classical argument used a continuous analog, given in the following lemma.

Lemma 7.2 *Let $f \in \mathcal{M}[I^n, \lambda]$ be any volume preserving homeomorphism of the unit cube I^n, $n \geq 2$, onto itself. Then given any $\epsilon > 0$, there exists another such homeomorphism $h \in \mathcal{M}[I^n, \lambda]$, $\|h - f\| < \epsilon$, such that for some arbitrarily fine dyadic decomposition $\{D_i\}_{i=1}^N$, there is a compact set $B \subset D_1$ with $\lambda(B) = \frac{1}{2}\lambda(D_1)$ and $h^i(B)$ in the interior of the dyadic cube D_{i+1}, for $i = 0, \ldots, N-1$.*

A version of this lemma was proved by Oxtoby and Ulam [88] by a long and elaborate argument which used Birkhoff's Individual Ergodic Theorem to find orbits of f which were approximately uniformly distributed with respect to the volume distribution. We will give a short proof based on modern ideas at the end of this chapter. We now show how generic ergodicity, Theorem 7.1, may be established based on this lemma, following the original approach of Oxtoby and Ulam. (Actually this proof shows that ergodic homeomorphisms are residual in $\mathcal{M}$ — G_δ-ness can be shown separately.)

Proof of Theorem 7.1 Let $\hat{D}_i$, $i = 1, 2, \ldots$, be an enumeration of all dyadic subcubes of I^n, of all orders. For indices i and j corresponding to dyadic cubes of the same order, define the subset $\mathcal{F}_{ij}$ of $\mathcal{M}[I^n, \lambda]$ by

$$h \in \mathcal{F}_{ij} \text{ if for some } h\text{-invariant set } A,$$

$$\lambda(A \cap \hat{D}_i) > \frac{3}{4}\lambda(\hat{D}_i) \text{ and } \lambda(A \cap \hat{D}_j) < \frac{1}{4}\lambda(\hat{D}_j).$$

Since for any measurable set $A \subset I^n$ with $0 < \lambda(A) < 1$ we can find such sets $\hat{D}_i$ and $\hat{D}_j$ in the definition of $\mathcal{F}_{ij}$, it follows from the definition of ergodicity that every nonergodic homeomorphism belongs to some set $\mathcal{F}_{ij}$, or $\mathcal{M}[I^n, \lambda] - \mathcal{E} \subset \bigcup_{ij} \mathcal{F}_{ij}$. In the previous expression, the union is taken over pairs i, j corresponding to dyadic cubes of the same order. So by Baire's Category Theorem it remains only to show that the sets $\mathcal{F}_{ij}$ are nowhere dense, in the uniform topology. Suppose that $f \in \mathcal{F}_{ij}$. Then by Lemma 7.2 there is a uniformly close homeomorphism $\tilde{h} \in \mathcal{M}[I^n, \lambda]$ which has a set B that is equidistributed, in the sense of that lemma, with respect to some dyadic decomposition $\{D_k\}_{k=1}^N$ which refines that of $\hat{D}_i$ and $\hat{D}_j$. Note that there is a uniform topology neighborhood of $\tilde{h}$, such that every h in this neighborhood has this same equidistribution property with the same set B. We claim that any h in this neighborhood of $\tilde{h}$ does not belong to $\mathcal{F}_{ij}$. Since this shows that every neighborhood of $f \in \mathcal{F}_{ij}$ contains a smaller neighborhood in the complement of $\mathcal{F}_{ij}$, then $\mathcal{F}_{ij}$ is nowhere dense in $\mathcal{M}[I^n, \lambda]$. To establish the claim suppose that A is any h-invariant set satisfying the condition

$$\lambda(A \cap \hat{D}_i) > \frac{3}{4}\lambda(\hat{D}_i).$$

Since A is h-invariant, it follows that $\lambda(A \cap h^k B) = c\lambda(D_k) = c\lambda(D_i)$, for some constant c, for $k = 0, \ldots, N-1$. Since the first N iterates of B under h fill up exactly half of $\hat{D}_i$ and of $\hat{D}_j$, it follows from the displayed inequality that $c > 1/4$. Consequently we have that

$$\lambda(A \cap \hat{D}_j) > \frac{1}{4}\lambda(\hat{D}_j),$$

and therefore we have established that h cannot belong to $\mathcal{F}_{ij}$. This shows that $\mathcal{E} \cap \mathcal{M}[I^n, \lambda]$ is a residual set, i.e., its complement is a subset of a countable union of nowhere dense sets. $\qquad\square$

As we said above, the proof of Lemma 7.2 given by Oxtoby and Ulam is long and complicated. We now give a short proof based on ideas developed earlier in the book.

Proof of Lemma 7.2 By Theorem 3.3 there is a cyclic dyadic permutation P of some dyadic decomposition $D_1, \ldots, D_N$ with $P(D_i) = D_{i+1}$, for $i = 1, \ldots, N-1$, and $P(D_N) = D_1$, with $\|Pf^{-1}\| < \epsilon$. Consequently, by Theorem 6.2, there is a homeomorphism $g \in \mathcal{M}[I^n, \lambda]$ with $\|g\| < \epsilon$, such that g is arbitrarily close to Pf^{-1} in the weak topology. Setting $h = gf$, this is equivalent to saying that there is a homeomorphism $h \in \mathcal{M}[I^n, \lambda]$, with $\|h - f\| < \epsilon$, which is arbitrarily close to P in the weak topology. Using an appropriate weak topology neighborhood, this gives a measurable set $\check{B}$ with $\lambda(\check{B}) > \frac{1}{2}\lambda(D_1)$, such that $h^i(\check{B})$ is a subset of D_{i+1}, $i = 0, \ldots, N-1$. Choosing a sufficiently large compact subset B of $\check{B}$ gives the condition of the lemma. $\qquad\square$

8

Uniform Approximation in $\mathcal{G}[I^n, \lambda]$ and Generic Properties in $\mathcal{M}[I^n, \lambda]$

8.1 Introduction

In this chapter we show that any volume preserving homeomorphism of the cube can be uniformly approximated by volume preserving automorphisms (not generally continuous) with certain specified measure theoretic properties. As shown in the previous section (when the property was ergodicity), this approximation can then be combined with the Lusin Theory to produce homeomorphisms possessing that property, and a version of Theorem C (of Chapter 1) for generic properties of volume preserving homeomorphisms of the cube.

Suppose we want to find homeomorphisms of I^n which have some particular measure theoretic property, such as ergodicity or weak mixing. Such a property can be designated by specifying a subset $\mathcal{V}$ of the space $\mathcal{G}[X, \mu]$ of all automorphisms of a Lebesgue space (X, μ), which we will take for convenience as all volume preserving bijections of (I^n, λ). We will only consider properties $\mathcal{V}$ which don't depend on the names of the points, i.e., which are conjugate invariant in $\mathcal{G}[I^n, \lambda]$. (This assumption means that $g \in \mathcal{V}$ implies $f^{-1}gf \in \mathcal{V}$ for all $f \in \mathcal{G}[I^n, \lambda]$.) In this context the statement at the beginning of this paragraph is equivalent to showing $\mathcal{V} \cap \mathcal{M}[I^n, \lambda]$ is nonempty. Many important measure theoretic properties in $\mathcal{G}[I^n, \lambda]$ are determined by a countable number of conditions, which each define an open set in the weak topology on $\mathcal{G}[I^n, \lambda]$, that is, they are G_δ subsets of $\mathcal{G}[I^n, \lambda]$. For such sets, we have already determined (in Corollary 6.3) a property which will ensure that $\mathcal{V} \cap \mathcal{M}[I^n, \lambda]$ is nonempty, namely that the uniform topology closure of $\mathcal{V}$ contains $\mathcal{M}[I^n, \lambda]$. In this chapter we give a simple property of $\mathcal{V}$ that ensures its uniform closure is all of $\mathcal{G}[I^n, \lambda]$, namely that $\mathcal{V}$ contains an antiperiodic automorphism. (An automorphism $g \in \mathcal{G}[I^n, \lambda]$ is called

53

antiperiodic if its periodic points have measure zero.) It is worth noting that if $\mathcal{V}$ is generic in $\mathcal{G}[I^n, \lambda]$ (with respect to the weak topology) then it must have a nonempty intersection with the generic property $\mathcal{E}$ (ergodic automorphisms). Since an ergodic automorphism is necessarily antiperiodic, $\mathcal{V}$ must contain an antiperiodic automorphism.

Since (for the present) we only need that the uniform closure of $\mathcal{V}$ contains $\mathcal{M}[I^n, \lambda]$ (rather than $\mathcal{G}[I^n, \lambda]$) that is all we shall prove in this chapter (Theorem 8.4). However, the more general result (with $\mathcal{G}[I^n, \lambda]$ rather than $\mathcal{M}[I^n, \lambda]$) is proved in Appendix 1 as Corollary A1.12 – this more general result is stated here as follows.

Theorem 8.1 *The conjugacy class of any antiperiodic automorphism is dense in* $\mathcal{G}[I^n, \lambda]$ *in the uniform topology. That is, given* $g, h \in \mathcal{G}[I^n, \lambda]$ *with* g *antiperiodic and* $\epsilon > 0$, *there is an* $f \in \mathcal{G}[I^n, \lambda]$ *satisfying* $\|f^{-1}gf - h\| < \epsilon$, *that is*

$$|f^{-1}gf(x) - h(x)| < \epsilon \quad \text{for } \lambda\text{-a.e. } x.$$

This result due to Alpern in 1979 ([10], and [11, Conjugacy Lemma]) is a strengthening of a similar result of Halmos ([70, Theorem 1] and Halmos's book [72]), that the conjugacy class of an antiperiodic transformation is dense in $\mathcal{G}[I^n, \lambda]$ in the *weak* topology. This result and Corollary 6.3 immediately give the following means of showing that certain measure theoretic properties not only exist but are in fact typical for volume preserving homeomorphisms of I^n. It generalizes the special cases where $\mathcal{V}$ represents ergodicity (established by Oxtoby and Ulam [88] – see previous chapter) or weak mixing (established by Katok and Stepin [76]) or in fact *any* property which is generic in $\mathcal{G}[I^n, \lambda]$. By combining the approximation result Theorem 8.1 with the measure preserving Lusin Theorem of Chapter 6, we will obtain the following simple condition for a property to be generic in $\mathcal{M}[I^n, \lambda]$. Since any property $\mathcal{V}$ which is generic in $\mathcal{G}[I^n, \lambda]$ contains an antiperiodic automorphism, the following result (ii) is the version of Theorem C (of Chapter 1) for the cube.

Theorem 8.2 *Let* $\mathcal{V}$ *be any conjugate invariant, weak topology* G_δ *subset of* $\mathcal{G}[I^n, \lambda]$ *which contains an antiperiodic automorphism. Then*

(i) $\mathcal{V}$ *is a dense* G_δ *subset of* $\mathcal{G}[I^n, \lambda]$ *in the weak topology, and*

(ii) $\mathcal{V} \cap \mathcal{M}[I^n, \lambda]$ *is a dense* G_δ *subset of* $\mathcal{M}[I^n, \lambda]$ *in the uniform topology.*

The result (ii) *remains true if* $\mathcal{M}[I^n, \lambda]$ *is replaced by* $\mathcal{M}[I^n, \partial I^n, \lambda]$ *(the volume preserving homeomorphisms fixing the boundary of the cube) in both instances.*

8.2 Rokhlin Towers and Stochastic Matrices

We leave the proof of Theorem 8.1 to Appendix 1 and instead prove a slightly weakened version of that theorem where the target automorphism h is actually a volume preserving *homeomorphism* – this is Theorem 8.4. To do this we will need to use either (i) Rokhlin's Tower Theorem (see Theorem 8.3) combined with the fixed point property of the cube (Brouwer's Theorem), or (ii) a stronger form of Rokhlin's Theorem in terms of stochastic matrices, without any topological properties. We will give both proofs. Then we will show how the main result of this part of the book, Theorem 8.2, follows from a weaker form of Theorem 8.1 (Theorem 8.4) and our earlier results in Chapter 6 on Continuous Approximation. The following result is stated for Lebesgue probability spaces (X, Σ, μ), although the reader may just think of it as the cube (I^n, λ) with volume measure, since measure theoretically these spaces are the same (via some measurable measure preserving conjugacy). See also Theorem A1.4.

Theorem 8.3 (Rokhlin Tower Theorem) *Let* (X, Σ, μ) *be a finite Lebesgue measure space and let* $g \in \mathcal{G}[X, \Sigma, \mu]$ *be an antiperiodic* μ-*preserving automorphism of* X. *Then given any positive integer* N *and any* γ, $0 < \gamma < \mu(X)$, *there is a measurable set* $A \in \Sigma$ *such that* $g^i A$ *are pairwise disjoint for* $i = 0, \ldots, N - 1$, *and*

$$\mu \left(\bigcup_{i=0}^{N-1} g^i A \right) = \gamma.$$

As a consequence of this purely measure theoretic theorem, we can now prove the following weakened form of Theorem 8.1, where the target automorphism h is restricted to homeomorphisms. This will be sufficient for our purposes, namely the proof of Theorem 8.2. Note that we later prove an extension of the above result, which we call the Multiple Tower Rokhlin Theorem.

Theorem 8.4 *Given any antiperiodic automorphism* g *in* $\mathcal{G}[I^n, \lambda]$, *any homeomorphism* h *in* $\mathcal{M}[I^n, \lambda]$ *and* $\epsilon > 0$, *there is an* $f \in \mathcal{G}[I^n, \lambda]$

satisfying $\|f^{-1}gf - h\| < \epsilon$, *i.e.*,

$$|f^{-1}gf(x) - h(x)| < \epsilon \quad for \ \lambda\text{-}a.e. \ x.$$

We will give two proofs of this result. The first proof [10] relies on the fact (Brouwer's Theorem) that the underlying space, the unit cube, has the fixed point property. The second proof [11] involves no topological properties, but rather uses a measure theoretic result that can be seen as a generalization of Rokhlin's Tower Theorem. The second proof is applicable to manifolds without the fixed point property.

Proof number one of Theorem 8.4 By our earlier result on cyclic dyadic approximation, Theorem 3.3, there is an appropriately indexed dyadic decomposition $\{\alpha_i\}_{i=1}^N$ such that the diameter of the sets $h(\alpha_i) \cup \alpha_{i+1}$, $i = 1, \ldots, N$ (where $\alpha_{N+1} = \alpha_1$) is less than $\epsilon/2$. Furthermore by Brouwer's Theorem one of the closed dyadic cubes, say α_N, contains a fixed point of h, so $\alpha_N \cap h(\alpha_N) \neq \emptyset$. Consequently

$$\text{diameter} \left[(h(\alpha_{N-1}) \cup \alpha_N) \cup (h(\alpha_N) \cup \alpha_1) \right] < \frac{\epsilon}{2} + \frac{\epsilon}{2} = \epsilon.$$

Now apply Rokhlin's Tower Theorem to the antiperiodic automorphism g to obtain a set A such that $g^{i-1}(A)$ are pairwise disjoint for $i = 1, \ldots, N - 1$, and $\lambda \left(\bigcup_{i=1}^{N-1} g^{i-1}(A) \right) = 1 - 1/N$. Let $f \in \mathcal{G}[I^n, \lambda]$ be any automorphism satisfying $f(\alpha_i) = g^{i-1}(A)$, $i = 1, \ldots, N - 1$, and $f(\alpha_N) = I^n - \bigcup_{i=1}^{N-1} g^{i-1}(A)$. Define $\hat{g} = f^{-1}gf$ and observe that $\hat{g}(\alpha_i) = \alpha_{i+1}$ for $i = 1, \ldots, N - 2$. Hence for $x \in \alpha_i$ with $i = 1, \ldots, N - 2$, we have $|h(x) - \hat{g}(x)| < \epsilon/2$, since the two points lie within the set $h(\alpha_i) \cup \alpha_{i+1}$. Next note that $\hat{g}(\alpha_{N-1} \cup \alpha_N) = \alpha_N \cup \alpha_1$. It follows that for $x \in \alpha_{N-1} \cup \alpha_N$, $\hat{g}(x)$ and $h(x)$ both belong to the set $(h(\alpha_{N-1}) \cup \alpha_N) \cup (h(\alpha_N) \cup \alpha_1)$. Since this set has diameter less than ϵ, the required estimate follows. $\qquad\square$

To prepare the second proof of Theorem 8.4, one that does not use the fixed point property of the underlying space (I^n), we begin with a few definitions. An $m \times m$ matrix $\mathbf{P}$ is called *stochastic* if its entries are nonnegative and its rows all sum to 1, and *doubly stochastic* if its columns also sum to 1. A stochastic matrix is called *mixing* if some power has all positive entries. The following result was proved in [11, Theorem 1]; a proof of a stronger version (for infinite matrices, see [23]) can be found in Appendix 1 as Theorem A1.9. A similar result for nonsingular transformations g is in [24].

Theorem 8.5 *Let $g \in \mathcal{G}[I^n, \lambda]$ be antiperiodic and let (p_{ij}) be an $m \times m$ mixing stochastic matrix. Then there is a measurable partition $\{\beta_i\}_{i=1}^m$ of I^n satisfying*

$$\frac{\lambda\left(g\left(\beta_i\right) \cap \beta_j\right)}{\lambda\left(\beta_i\right)} = p_{ij} \qquad \text{for all } i, j = 1, \dots, m.$$

We now give our second proof of Theorem 8.4, one that does not use the fixed point property.

Proof number two of Theorem 8.4 Begin by choosing a dyadic decomposition $\{\alpha_i\}_{i=1}^m$ of I^n sufficiently fine so that $2|\alpha_i| + |h\left(\alpha_i\right)| < \epsilon$, for $i = 1, \dots, m$ (where $|S|$ denotes the diameter of a set S). Number the α_i cyclically so that α_i is adjacent to α_{i+1} for $i < m$, and α_m is adjacent to α_1. Define $\mathbf{A}$ to be the $m \times m$ doubly stochastic matrix given by $a_{ij} = m\,\lambda\left(h\left(\alpha_i\right) \cap \alpha_j\right)$. Define $\mathbf{B}$ to be the $m \times m$ doubly stochastic matrix given by

$$b_{ij} = \begin{cases} 1/2 & \text{if } i = j \text{ or } i = j - 1 \bmod(m), \\ 0 & \text{otherwise.} \end{cases}$$

Then it is easy to see that $\mathbf{P} = \mathbf{AB}$ is doubly stochastic and mixing (in fact $\mathbf{P}^m > 0$). Now apply the above theorem to the antiperiodic automorphism g and the mixing stochastic matrix $\mathbf{P}$ with entries p_{ij}, to obtain a partition $\{\beta_i\}_{i=1}^m$ satisfying

$$\frac{\lambda\left(g\left(\beta_i\right) \cap \beta_j\right)}{\lambda\left(\beta_i\right)} = p_{ij} \qquad \text{for all } i, j = 1, \dots, m.$$

Note in particular that $\lambda(g\left(\beta_i\right) \cap \beta_j) > 0$ only if $p_{ij} > 0$. Since both the β and α partitions divide I^n into sets of volume $1/m$, there is a automorphism $f \in \mathcal{G}[I^n, \lambda]$ for which $f\left(\alpha_i\right) = \beta_i$, $i = 1, \dots, m$. Any point $x \in I^n$ belongs to some set α_i and its image $f^{-1}gf(x)$ belongs to some set α_k. Since $f(x)$ belongs to β_i, it follows that $gf(x)$ belongs to some set β_k for which $p_{ik} > 0$, except possibly for a set of x's of measure zero. If $p_{ik} > 0$ then for some j, a_{ij} and b_{jk} are both positive. Since b_{jk} is positive, either $j = k - 1 \bmod (m)$, or $j = k$. Similarly, since a_{ij} is positive we have $h\left(\alpha_i\right) \cap \alpha_j \neq \emptyset$. Thus the point $h(x)$ belongs to a set $h\left(\alpha_i\right)$ which intersects a set α_j which is adjacent to the set α_k containing $f^{-1}gf(x)$. It follows that

$$|h(x) - f^{-1}gf(x)| \leq |h\left(\alpha_i\right)| + |\alpha_j| + |\alpha_k| < \epsilon$$

for λ-a.e. x, by the initial choice of the size of the α partition. $\qquad \square$

Note that the only place in the above proof that the continuity of h was used (an assumption of Theorem 8.4 but not Theorem 8.1) was in the initial choice of the partition α_i with small diameter and *hence* small diameter of $h(\alpha_i)$. If h is only assumed to belong to $\mathcal{G}[I^n, \lambda]$ then we can still obtain such a partition, but we can no longer guarantee that all of its elements have the same volume. A similar but more complicated proof then proceeds as above but with stochastic matrices rather than doubly stochastic matrices. The details are given in [11]. In any case, we can now proceed to prove our main result on generic properties of volume preserving homeomorphisms of the cube.

Proof of Theorem 8.2   Since $\mathcal{V}$ is conjugate invariant and contains an antiperiodic automorphism, Theorem 8.4 says that the uniform topology closure of $\mathcal{V}$ contains $\mathcal{M}[I^n, \lambda]$. It then follows from Corollary 6.3 that (ii) $\mathcal{V} \cap \mathcal{M}[I^n, \lambda]$ is a dense G_δ subset of $\mathcal{M}[I^n, \lambda]$ in the uniform topology. Part (i) then follows from the fact that the uniform topology is finer than the weak topology and the fact (Theorem 6.2) that $\mathcal{M}[I^n, \lambda]$ is dense in $\mathcal{G}[I^n, \lambda]$ in the weak topology. $\square$

Part II

Measure Preserving Homeomorphisms of a
Compact Manifold

9

Measures on Compact Manifolds

9.1 Introduction to Part II

Up to now we have restricted our attention to volume preserving homeomorphisms of the cube, and have proved a number of results for this space $\mathcal{M}[I^n, \lambda]$. In this part of the book (Chapters 9 and 10) we show how the results already obtained for $\mathcal{M}[I^n, \lambda]$ apply more generally to the space $\mathcal{M}[X, \mu]$ whenever X is any compact connected manifold (we allow situations where our manifold X could possibly have nonempty boundary as for example when $X = I^n$) and μ belongs to a certain class of finite measures. In other words, we will show that there was really no loss of generality in restricting our attention to the cube with volume measure, where the intuition was clearer.

We note for later purposes that the situation is very different for noncompact manifolds, in that results obtained for the 'standard noncompact manifold' R^n do *not* go over unchanged to arbitrary noncompact manifolds. That is, for compact manifolds the topological type of the manifold is irrelevant, but for noncompact manifolds the end structure is important. But these are matters to be dealt with in Part III.

9.2 General Measures on the Cube

We begin our analysis by retaining for the moment the cube I^n, $n \geq 2$, as our manifold, but now endowing it with a more general Borel probability measure μ. Our first question concerns the assumptions we have to make about the measure μ to ensure that our main genericity result from Part I (Theorem 8.2) continues to hold for the space $\mathcal{M}[I^n, \mu]$. Recall that this result says that any typical measure theoretic property $\mathcal{V}$, such as ergodicity or weak mixing, is also typical in $\mathcal{M}[I^n, \lambda]$.

We begin by considering certain possible properties of a measure μ which must be ruled out. For this discussion let f be a homeomorphism in $\mathcal{M}[I^n, \mu]$.

(i) *atoms:* If μ has atoms (points of positive measure), then those with any fixed measure form an invariant set, so for f to be ergodic these points must be cyclically permuted. But such an f is not weak mixing, and so (taking $\mathcal{V}$ to be the weak mixing automorphisms) Theorem 8.2 is not true in $\mathcal{M}[I^n, \mu]$ for such a measure μ.

(ii) *open set with zero measure:* Here we only give an example to show that this might create a problem. Suppose that μ assigns zero measure to the open subset of the square where the first coordinate lies in the interval $(1/2, 3/4)$, and μ equals $4/3$ times area measure (λ) on its complement. Then any $f \in \mathcal{M}[I^2, \mu]$ which fixes the boundary cannot be ergodic, as the sets to the left and right of this vertical strip are each invariant.

(iii) *boundary has positive measure:* By the property called 'invariance of domain', ∂I^n is invariant under f, so if $0 < \mu(\partial I^n) < 1$ then f cannot be ergodic.

We exclude these problems by defining what we call (in honor of Oxtoby and Ulam) an OU measure.

Definition A Borel measure μ on a manifold X is called an *OU measure* if it is

(i) *nonatomic* (this means it is zero on singleton sets – has no 'atoms'),
(ii) *locally positive* (this means it is positive on every nonempty open set), and
(iii) *zero on* the manifold *boundary*.

It is worth recalling that every Borel measure is *locally finite*, that is, finite on compact sets. Thus an OU measure on a compact manifold is finite, and hence will often be normalized to a probability measure (with $\mu(X) = 1$). Similarly (though this will not be of any significance until Part III) an OU measure on a sigma compact manifold is sigma finite.

An important though simple observation is that Lebesgue measure λ on the cube I^n is an OU probability measure. Almost as simple is the observation that any measure ν on I^n which is *homeomorphic to* λ (that is, of the form $\nu(A) = \lambda(h(A))$, for some homeomorphism h of I^n and all measurable sets A) is also an OU measure. In this case we write $\nu = \lambda h$.

The following converse of this statement is also true, and has been an important tool in the study of measure preserving homeomorphisms.

Theorem 9.1 (Homeomorphic Measures Theorem) *A Borel probability measure μ on the n-cube I^n is homeomorphic to Lebesgue measure λ on I^n if and only if it is an OU probability measure. In other words, there is a homeomorphism $h : I^n \to I^n$ with $\mu = \lambda h$ if and only if the Borel probability measure μ is nonatomic, locally positive, and has $\mu(\partial I^n) = 0$. Furthermore, given any homeomorphism $g : I^n \to I^n$, we can choose h to equal g on ∂I^n; in particular, we can take h to be the identity on the boundary.*

This result was conjectured by Ulam in 1936 and a proof was obtained at that time by J. von Neumann but was not published (although a handwritten manuscript of that proof is cited and reviewed in [105, vol. II, p. 558]). The first published proof is by Oxtoby and Ulam (Theorem 2, p. 886 of [88]), who used Baire category methods. We will give a version of their proof in Appendix 2. Another proof of this result due to Goffman and Pedrick can be found in [66]. The latter result is elaborated upon and strengthened in Chapter 7 of the 1998 book by Goffman, Nishiura and Waterman (see Theorem 7.1, p. 90, [65]), to prove a homotopy version of the Homeomorphic Measures Theorem. An infinite dimensional analog of the homeomorphic measures theorem for the Hilbert cube, I^∞, has been obtained by Oxtoby and the second author in [92] (an unpublished proof has also been obtained by B. Weiss [106]). The extension of the result to Hilbert cube manifolds is due to Prasad [97].

The reader may have noticed the absence in the above theorem of our usual caveat that $n \geq 2$. This is because the result is true even in dimension $n = 1$. In fact this is the only easy case; simply define $h(x) = \mu[0, x)$ for $x \in [0, 1]$. However, in higher dimensions the theorem becomes harder. What makes this theorem seem implausible in higher dimensions is that subsets of dimension strictly less than n can support some or all of the measure μ. For example, if μ_1 is any nonatomic Borel probability measure on I^n with support on some line segment L (say uniform measure), then the measure $\mu = (\mu_1 + \lambda)/2$ is an OU probability measure, and so must be of the form λh for some homeomorphism h of I^n. This means that the arc $h(L)$ must be a set of planar (Lebesgue) measure $1/2$! Another example of an OU probability measure is given by the following construction: For $i = 1, 2, \ldots$, let C_i be disjoint Cantor subsets of the interior of I^n whose union is dense in I^n, and let μ_i be arbitrary nonatomic Borel probability measures with supports on C_i. Then $\mu(U) = \sum_{i=1}^{\infty} \mu_i(U \cap C_i)/2^i$ is an OU probability measure.

The Homeomorphic Measures Theorem immediately extends our genericity result (Theorem 8.2) for $\mathcal{M}[I^n, \lambda]$ to the space $\mathcal{M}[I^n, \mu]$ for any OU measure μ. The idea is to normalize μ (assume that $\mu(I^n) = 1$) and observe that if $\mu = \lambda h$ then h establishes a homeomorphism between the two function spaces (from $\mathcal{M}[I^n, \lambda]$ to $\mathcal{M}[X, \mu]$) by the correspondence $f \leftrightarrow h^{-1}fh$. The following result may be thought of as a half-way house between the genericity result (Theorem 8.2) for $\mathcal{M}[I^n, \lambda]$ and for $\mathcal{M}[X, \mu]$ (where X is a compact manifold) given in Corollary 10.4.

Theorem 9.2 *Let μ be an OU probability measure in I^n, $n \geq 2$. Let $\mathcal{V}$ be a conjugate invariant, weak topology G_δ subset of $\mathcal{G}[I^n, \mu]$, which contains an antiperiodic automorphism. Then $\mathcal{V} \cap \mathcal{M}[I^n, \mu]$ is a dense G_δ subset of $\mathcal{M}[I^n, \mu]$ with respect to the uniform topology.*

Proof By the Homeomorphic Measures Theorem, there is a homeomorphism $h : I^n \to I^n$ such that $\mu = \lambda h$, where λ denotes Lebesgue measure. First note that if $\mathcal{V} \subset \mathcal{G}[I^n, \mu]$, then $\tilde{\mathcal{V}} = h\mathcal{V}h^{-1} \subset \mathcal{G}[I^n, \lambda]$ as the following easy calculation demonstrates: when $g \in \mathcal{G}[I^n, \mu]$, then $\tilde{g} = hgh^{-1} \in h\mathcal{V}h^{-1}$ and

$$
\begin{aligned}
\lambda \tilde{g}(A) &= \lambda h g h^{-1}(A) \\
&= \mu g h^{-1}(A) \\
&= \mu h^{-1}(A) \\
&= \lambda(A)
\end{aligned}
$$

shows $\tilde{g} \in \mathcal{G}[I^n, \lambda]$. Furthermore note that in the above, if $g \in \mathcal{M}[I^n, \mu]$, then $\tilde{g} = hgh^{-1} \in \mathcal{M}[I^n, \lambda]$.

Therefore we can apply Theorem 8.2 to the subset $h\mathcal{V}h^{-1}$ of $\mathcal{G}[I^n, \lambda]$, to assert that $h\mathcal{V}h^{-1} \cap \mathcal{M}[I^n, \lambda]$ is dense and G_δ in $\mathcal{M}[I^n, \lambda]$ with respect to the uniform topology. It follows that $h^{-1}\left(h\mathcal{V}h^{-1} \cap \mathcal{M}[I^n, \lambda]\right)h = \mathcal{V} \cap \mathcal{M}[I^n, \mu]$ is dense and G_δ in $\mathcal{M}[I^n, \mu]$, as required. $\square$

9.3 Manifolds

Up to this point in the book the usual domain, or phase space, that we have considered is the unit cube I^n. From this point on, we will take our domain to be a manifold X, possibly with a manifold boundary (denoted by ∂X). The dimension of the manifold will be denoted n, and assumed to be at least 2. In this part of the book we will assume that X is compact, while in Part III we will weaken this assumption to sigma compactness. The manifold X will always be endowed with a measure μ, which is assumed to be an OU measure, as defined in the previous section. The weak and uniform topologies have been defined on $\mathcal{G}[X, \mu]$ (and hence the subset $\mathcal{M}[X, \mu]$) in Chapters 1 and 2. Recall that the definition of the uniform topology requires that X is equipped with a metric d. It will be useful to assume that the metric d maintains a property possessed by the Euclidean and maximum metrics on the cube, that was used in Part I:

Given points $x, y \in X$, with $d(x, y) < \delta$, there is an arc $L : [0, \delta] \to X$ with $L(0) = x$, $L(\delta) = y$ and having Lipshitz constant 1.

Such a metric exists for example when X is a simplicial complex in some Euclidean space. We thank Ethan Akin for this useful observation. This property was used in Theorem 2.4 (second line of the proof) and indirectly in Theorem 6.2. This property is not essential, but results in a simpler statement of the norm preservation part of the Lusin Theorem.

Actually, we have already briefly considered manifolds other than I^n, namely the n-torus and the annulus, in our discussion of fixed points in Chapter 5. We were able to define dyadic permutations of these manifolds because we could view them simply as the unit cube with certain boundary identifications. This identification also enabled us to define volume on these manifolds. The topological aspect of the boundary identification idea is well known to be applicable to an arbitrary compact manifold. For this purpose we will use the strictly topological result of M. Brown stated below (for a nice proof of this see [52]). In the next section we will use the Homeomorphic Measures Theorem to extend Brown's result to additionally obtain an arbitrary OU probability measure μ from Lebesgue measure on the cube via such a boundary identification.

Theorem 9.3 (M. Brown [45]) *For any compact connected n-dimensional manifold X where $n \geq 2$, there exists a continuous map $\phi : I^n \to X$ such that*

(i) *ϕ maps onto X*

(ii) *$\phi|\operatorname{Int} I^n$ is a homeomorphism of the interior of I^n onto its image*

(iii) *$\phi(\partial(I^n))$ is a closed nowhere dense set, disjoint from $\phi(\operatorname{Int} I^n)$*

Whenever ϕ is a map from I^n to X satisfying (i)–(iii), we shall call the set $\phi(\partial I^n)$ in X a set of *singular points* of X and refer to ϕ as a *topological Brown map*. Note that in this definition of the singular points of X, a different map ϕ will give a different set of singular points. Nevertheless, the proof of Brown's theorem actually shows that if X has any boundary, then the boundary always consists of singular points. For example if X is the torus T^2 then the union of the great circles in each coordinate direction would form a set of singular points of T^2. When X is the 2-dimensional annulus, the singular points include the two bounding circles, together with a line segment connecting them. In general, the set of singular points consists of the manifold boundary together with a 'scar' on X so that if X is cut open along the scar, then 'what is left' $(X - \phi(\partial(I^n)))$ is homeomorphic to the interior of I^n.

Given an OU probability measure μ on X, we would like to be able to say that 'what is left' is not only homeomorphic to the interior of I^n, but also has full measure, that is

$$\mu\left(\phi(\partial(I^n))\right) = 0.$$

In fact, if we can ensure this, we can use the Homeomorphic Measures Theorem to show that the Brown map can be chosen to take Lebesgue measure λ into the given measure μ. In the next section we will show that there is some homeomorphism of I^n such that a given closed nowhere dense set (and therefore any set of singular points) can be 'moved by the homeomorphism' to a set of zero measure. That result will then allow us to choose a Brown map so that the singular points have zero measure.

9.4 Measures on Compact Manifolds

In the previous section we stated Brown's Theorem, which enables us to represent any compact manifold X as the unit cube I^n with appropriate boundary identifications. This is done via the map $\phi : I^n \to X$. In this section we will extend Brown's Theorem to show further that if μ is any OU probability measure on the manifold X, then ϕ can be chosen such that it takes Lebesgue measure λ on the cube into the measure μ. In order to do this, we must first establish the weaker result that we can choose ϕ such that the 'scar' of singular points has μ-measure zero, that is, $\mu\left(\phi(\partial(I^n))\right) = 0$. Since according to Brown's Theorem the closed set $\phi(\partial(I^n))$ is *nowhere dense*, we rely on a result of Oxtoby and Ulam [88] (Theorem 9.5 below) that a nowhere dense set can be mapped homeomorphically into a set of zero measure. But first we prove a lemma in which 'zero measure' is weakened to 'small measure'.

Lemma 9.4 *Suppose μ is measure on I^n such that $\mu(\partial I^n) = 0$ and F is a closed nowhere dense subset of I^n. Then for each δ and $\epsilon > 0$, there exists a homeomorphism $h \in \mathcal{H}[I^n, \partial I^n]$, with $\|h\| < \delta$ and $\mu h(F) < \epsilon$.*

Proof Consider for each $t \in [0, 1]$ the following section in the first co-ordinate, $\{x = (x_1, \ldots, x_n) : x_1 = t\}$. Since these are all disjoint for different t's, there can be at most countably many such sections with positive measure. Similarly there are only countably many sections in any coordinate with positive measure. So we may avoid those sections in choosing hyperplanes to form a finite subdivision $D_1, \ldots, D_k$ of I^n into rectangles such that $\text{diam}(D_i) < \delta$ and $\mu(\partial D_i) = 0$ for each i. Let U be an open neighborhood of $\bigcup_{i=1}^{k} \partial D_i$ such that $\mu(U) < \epsilon$. Because the given set F has no interior, let $D_i' \subset \text{Int} \, D_i$ be a small subrectangle with faces parallel to the faces of D_i and such that $F \cap D_i' = \emptyset$. Then for each i let $h_i \in \mathcal{H}[D_i, \partial D_i]$ be a homeomorphism of the cube D_i which maps the annular region $D_i - D_i'$ into $U \cap D_i$. The homeomorphism h of I^n obtained by piecing together the h_i's has the required property. $\quad\square$

We now strengthen this result from 'small measure' to 'zero measure', using a Baire category proof of Oxtoby and Ulam [88] which shows that in fact *most* (topologically speaking) homeomorphisms take the set F into a set of zero measure. This result establishes a connection between topological smallness and measure theoretic smallness. For more connections of this type, the reader is again referred to Oxtoby's elegant book [91].

Theorem 9.5 *Suppose that X is a compact connected n-manifold and μ is a Borel measure on X which vanishes on ∂X. Let A be a closed nowhere dense subset of X. Then $\{h \in \mathcal{H}[X] : \mu h(A) = 0\}$ is a dense G_δ subset of $\mathcal{H}[X]$.*

Proof Each point $x \in X$ has a neighborhood B' homeomorphic to I^n (say by homeomorphism $\psi : B' \to I^n$). Since there are uncountably many concentric subcubes of I^n around $\psi(x)$, and at most countably many of these subcubes give rise to neighborhoods of x whose boundaries have positive μ measure, let $B \subset B'$ be a neighborhood of x, homeomorphic to I^n with $\mu(\partial B) = 0$. Since X is compact, let $B_1, \ldots, B_k$ be a covering of X by sets homeomorphic to I^n such that $\mu(\partial B_i) = 0$ for each i. For each $i = 1, \ldots, k$ and each positive integer j, define

$$\mathcal{H}_{ij} = \{h \in \mathcal{H}[X] : \mu(h(A) \cap B_i) < 1/j\}.$$

Clearly

$$\{h \in \mathcal{H}[X] : \mu h(A) = 0\} = \bigcap_{ij} \mathcal{H}_{ij}.$$

The technique employed in the previous lemma, applied to B_i, establishes that each $\mathcal{H}_{ij}$ is dense in $\mathcal{H}[X]$.

$\mathcal{H}_{ij}$ is clearly open in the weak topology, and consequently open in the finer uniform topology.   □

We are now in a position to state and prove the measure preserving extension of Brown's Theorem that was discussed at the beginning of this section.

Theorem 9.6 *Let μ be an OU probability measure on a compact connected n-dimensional manifold X, $n \geq 2$. There exists a map $\phi : I^n \to X$ such that*

   (i) *ϕ maps onto X*
   (ii) *$\phi|\operatorname{Int} I^n$ is a homeomorphism of the interior of I^n onto its image*
   (iii) *$\phi(\partial(I^n))$ is a closed nowhere dense set, disjoint from $\phi(\operatorname{Int} I^n)$*
   (iv) *$\mu(\phi(\partial I^n)) = 0$*
   (v) *ϕ is a measure preserving map of $(\operatorname{Int} I^n, \lambda)$ onto its image; i.e.,*
       $$\mu(\phi(U)) = \lambda(U) \text{ for all Borel sets } U \subset \operatorname{Int} I^n.$$

Proof The first three parts constitute Brown's Theorem. So let ϕ_1 be any map satisfying (i)–(iii). We show how to modify it to obtain (iv) and (v) as well. By the previous theorem, since $\phi_1(\partial I^n)$ is a closed nowhere dense set in X, there is a homeomorphism h of X such that $h\phi_1$ satisfies (i)–(iv).

Define $\nu = \mu h\phi_1$. Then ν is an OU probability measure on I^n. Thus the Homeomorphic Measures Theorem (9.1) ensures that there is a homeomorphism $f \in \mathcal{H}[I^n, \partial I^n]$ so that $\nu = \lambda f$. Thus $\mu h\phi_1 f^{-1} = \lambda$, and so $\phi = h\phi_1 f^{-1}$ is the required measure preserving map satisfying (i)–(v).   □

Any map ϕ satisfying conditions (i)–(v) will be called a *Brown map*. Note that in addition to being a topological Brown map from I^n to X, ϕ also gives a measure theoretic isomorphism between the measure spaces (I^n, λ) and (X, μ). Note that a measure preserving map is an isomorphism of two measure spaces if it is a measure preserving bijection 'modulo sets of measure zero'. In this situation, although ϕ is not necessarily a measure preserving bijection from (I^n, λ) to (X, μ), ϕ is a measure preserving bijection once we remove the λ-null set ∂I^n from the domain, and its image $\phi(\partial I^n)$ (which has μ-measure zero) from the range (X,μ).

9.5 Typical Properties in $\mathcal{M}[X,\mu]$

We can use our measure preserving version of Brown's Theorem (Theorem 9.6) to show that any typical measure theoretic property can be represented by a homeomorphism in $\mathcal{M}[X,\mu]$ for any compact manifold. This gives a partial extension of Theorem 8.2, bootstrapping it from the domain (I^n,λ) to (X,μ). The full extension of Theorem 8.2 to the domain (X,μ) will be given in the next chapter, and cannot be obtained by the methods used here.

Theorem 9.7 *Let* $\mathcal{V}$ *be a conjugate invariant weak topology G_δ subset of* $\mathcal{G}[X,\mu]$ *which contains an antiperiodic automorphism. Fix any Brown map* $\phi : I^n \to X$, *and let* $K = \phi(\partial I^n)$. *Then* $\mathcal{V} \cap \mathcal{M}[X,K,\mu]$ *is a uniform topology dense G_δ subset of the space* $\mathcal{M}[X,K,\mu]$, *that is, of the subset of* $\mathcal{M}[X,\mu]$ *consisting of μ-preserving homeomorphisms of X equal to the identity on K.*

Proof Since the map ϕ is a measure theoretic isomorphism (measure preserving bijection between sets of full measure) between the Lebesgue probability spaces (I^n,λ) and (X,μ), and $\lambda = \mu\phi$, it follows that $\phi^{-1}\mathcal{V}\phi$ is a conjugate invariant weak topology G_δ subset of $\mathcal{G}[I^n,\lambda]$ which contains an antiperiodic automorphism. Denote by $\mathcal{M}[I^n,\partial I^n,\lambda]$ the subset of volume preserving homeomorphisms of I^n fixing the boundary ∂I^n. It follows from the last sentence in Theorem 8.2 that $\phi^{-1}\mathcal{V}\phi \cap \mathcal{M}[I^n,\partial I^n,\lambda]$ is a dense G_δ subset of $\mathcal{M}[I^n,\partial I^n,\lambda]$ with respect to the uniform topology. Next observe that ϕ induces a homeomorphism T from $\mathcal{M}[I^n,\partial I^n,\lambda]$ to $\mathcal{M}[X,\phi(\partial I^n),\mu]$ by the formula $T(h) = \phi h\phi^{-1}$. It follows that $T\left(\phi^{-1}\mathcal{V}\phi \cap \mathcal{M}[I^n,\partial I^n,\lambda]\right) = \mathcal{V}\cap\mathcal{M}[X,K,\mu]$ is a dense G_δ subset of $\mathcal{M}[X,\phi(\partial I^n),\mu]$ in the uniform topology. $\qquad\square$

It is worth emphasizing the fact that the bootstrapping technique (of using our version of Brown's Theorem to extend results from (I^n,λ) to (X,μ)) used above *cannot* be adapted to prove that sets are dense in the full space $\mathcal{M}[X,\mu]$. In order to obtain a version of Theorem 8.2 valid in the full space $\mathcal{M}[X,\mu]$ we will have to wait until the next chapter, where we prove a norm bounded Lusin theorem for the full space. The (very) alert reader will recall that one of our proofs of Theorem 8.4 (and hence indirectly, of Theorem 8.2) used the fixed point property of the cube. Since in general the compact manifold X does not have this property, a bootstrap proof (as above) for the manifold could not be expected to work.

10

Dynamics on Compact Manifolds

10.1 Introduction

In the previous chapter, we defined the space $\mathcal{M}[X, \mu]$ of all homeomorphisms of a compact connected manifold X which preserve an OU probability measure μ. In addition, we proved the existence of a 'Brown map' $\phi : I^n \to X$, and used it to prove (Theorem 9.7) that typical measure theoretic properties $\mathcal{V}$ are also typical in the subspace $\mathcal{M}[X, \phi(\partial I^n), \mu]$ of $\mathcal{M}[X, \mu]$ consisting of homeomorphisms which pointwise fix the singular set $K = \phi(\partial I^n)$. In the next section of this chapter we will show (Theorem 10.3) that this genericity result holds for the full space $\mathcal{M}[X, \mu]$, although it cannot be established by simple bootstrapping arguments involving the Brown map.

The final section of this chapter considers the existence of fixed points for volume preserving homeomorphisms of the open unit n-cube. Recall that we proved earlier (Theorem 5.5) Montgomery's observation that for $n = 2$ all such homeomorphisms which are orientation preserving have a fixed point. We will negatively answer the question of Bourgin as to whether Montgomery's result can be extended to higher dimensions or to orientation reversing homeomorphisms. The main tool will be the Homeomorphic Measures Theorem (Theorem 9.1), stated in the previous chapter.

10.2 Genericity Results for Manifolds

One of the main results proved in Part I of this book showed which measure theoretic properties are generic (in the sense of Baire category) for volume preserving homeomorphisms of the cube I^n – we saw that these properties include any (conjugate invariant) property which is generic for

71

the volume preserving automorphisms. In order to extend this genericity result (Theorem 8.2) from $\mathcal{M}[I^n, \lambda]$ to $\mathcal{M}[X, \mu]$, we must first extend the two theorems on which it is based: the Conjugacy Approximation Theorem (Theorem 8.1) and the Measure Preserving Lusin Theorem (Theorem 6.2). In extending these results to the context of $\mathcal{M}[X, \mu]$ it will be convenient to fix a particular Brown map $\phi : I^n \to X$ and denote the corresponding set of singular points $\phi(\partial I^n)$ as K. This map enables us to define a *quasi-dyadic cube* in X as the image under ϕ of an open dyadic cube in I^n.

Our extended version of the conjugacy result is the following.

Theorem 10.1 (Conjugacy Approximation Theorem) *Let μ be an OU probability measure on a compact connected n-manifold X, $n \geq 2$, with a given metric d. Given $\epsilon > 0$ and $h, g \in \mathcal{G}[X, \mu]$, where g is antiperiodic, there is an $f \in \mathcal{G}[X, \mu]$ with*

$$\|fgf^{-1} - h\| \equiv \operatorname{ess\,sup}_{x \in X} \ d(fgf^{-1}(x), h(x)) < \epsilon.$$

Proof There are two ways of proving this result, depending on whether we are willing to use the countable version of the Multiple Tower Rokhlin Theorem (Theorem A1.4 in Appendix 1), or the simpler finite version (or rather its consequence) Theorem 8.5. In the former case, we simply observe that this result is a special case of Corollary A1.12. In the latter case (this is the historical case, as this theorem was first proved [11] before the infinite version of the Tower Theorem) we can simply copy the second proof (the one not using the fixed point property of the cube) of Theorem 8.4, substituting quasi-dyadic cubes for dyadic cubes. If we wish to have the form of Theorem 8.1 where h is not assumed to be a homeomorphism, we need to also incorporate the remarks immediately after the second proof of Theorem 8.4.   □

We now prove the following general version of the Measure Preserving Lusin Theorem (6.2(i)).

Theorem 10.2 (Measure Preserving Lusin Theorem) *Let μ be an OU probability measure on the compact manifold X with metric d. Given $g \in \mathcal{G}[X, \mu]$ with $\|g\| \equiv \operatorname{ess\,sup}_{x \in X} d(x, g(x)) < \epsilon$, and a weak topology neighborhood $\mathcal{W}$ of g, there is a μ-preserving homeomorphism $h \in \mathcal{M}[X, \mu]$, which fixes the manifold boundary of X, and satisfies*

$$h \in \mathcal{W} \text{ and } \|h\| < \epsilon.$$

Proof By Brown's Theorem (Theorem 9.6) we may represent (X, μ) as (I^n, λ) with possible boundary identifications. This enables us to define dyadic cubes on X. The proof of Theorem 6.2(i) can now be carried out on (X, μ). The only modification of that proof to the manifold setting is that in the final **step 3** which uses Lemma 6.6 we must use the hypothesis given in Section 9.3 regarding the metric d in order to justify the application of Theorem 2.4. $\qquad\square$

It is instructive to see why the above result fails if we do not make the special assumption on the metric d mentioned in Section 9.3 (points a distance less than δ apart can be connected by an arc of length less than δ). Suppose that our manifold X is obtained by removing a thin vertical rectangle from the middle of the bottom of the square $I^2 = [0, 1]^2$, and keeping the Euclidean metric d. Specifically, define

$$X = I^2 - (1/2 - \epsilon/4, 1/2 + \epsilon/4) \times [0, 1/2).$$

Consider the automorphism $g \in \mathcal{G}[X, \lambda]$ which transposes (by translation for example) the two squares $S^- = (1/2 - \epsilon/2, 1/2 - \epsilon/4) \times (0, \epsilon/4)$ and $S^+ = (1/2 + \epsilon/4, 1/2 + \epsilon/2) \times (0, \epsilon/4)$ which lie at the bottom of X on either side of the missing vertical strip. Off $S^- \cup S^+$, g is the identity. With respect to the Euclidean metric d, we have $\|g\| < \epsilon$. If a homeomorphism $h \in \mathcal{M}[X, \lambda]$ is sufficiently close to g in the weak topology, then it will map some point x in S^- into S^+ and some point y just to the left of S^- very close to itself. If L is any small arc connecting x to y, then it contains a point z with $h(z) \in \{1/2\} \times [1/2, 1]$. Consequently the distance $d(z, h(z))$ is not much smaller than $1/2$, and we cannot have $\|h\| < \epsilon$.

On the other hand, if we did not need the norm bound $\|h\| < \epsilon$ in the Lusin Theorem above, we could give a simple proof bootstrapping from the version for the cube (without the norm bound) and the Brown map ϕ: Simply approximate $\phi^{-1} g \phi$ by a homeomorphism $\hat{h} \in \phi^{-1} \mathcal{W} \phi \cap \mathcal{M}[I^n, \partial I^n]$ and define $h = \phi \, \hat{h} \phi^{-1}$. But of course we do need the norm bound in our Lusin theorem, so we must use the proof given above, in which the approximations are carried out in the space X.

We can now state our main theorem on generic properties of measure preserving homeomorphisms of a compact manifold. This extension of the similar result for the unit cube (Theorem 8.2) follows from the manifold Theorems 10.2 and 10.1 in the same way as Theorem 8.2 followed from the cube Theorems 6.2 and 8.1. (The two required theorems are the

Measure Preserving Lusin Theorem and the Conjugacy Approximation Theorem.)

Theorem 10.3 *Let μ be an OU measure on a compact connected manifold X. Let $\mathcal{V}$ be any conjugate invariant, weak topology G_δ subset of $\mathcal{G}[X,\mu]$ which contains an antiperiodic transformation. Then*

(i) *$\mathcal{V}$ is a dense G_δ subset of $\mathcal{G}[X,\mu]$ in the weak topology, and*

(ii) *$\mathcal{V} \cap \mathcal{M}[X,\mu]$ is a dense G_δ subset of $\mathcal{M}[X,\mu]$ in the uniform topology.*

The above result gives a sufficient condition for a measure theoretic property (the set $\mathcal{V}$) to be 'typical' in both $\mathcal{G}[X,\mu]$ and $\mathcal{M}[X,\mu]$, provided that we interpret 'typical' to be 'dense G_δ' in the appropriate space. However, since many properties were first proved to be typical in the easier measure theoretic context ($\mathcal{G}[X,\mu]$) and only later for homeomorphisms ($\mathcal{M}[X,\mu]$) (e.g., weak mixing by Halmos [70] in 1944 and Katok and Stepin [76] in 1970) it is instructive to see how the above theorem can be rephrased to say the following: *If a measure theoretic property is typical in the purely measure theoretic context, it is also generic for measure preserving homeomorphisms.* If we interpret a 'measure theoretic property' to be a conjugate invariant subset of $\mathcal{G}[X,\mu]$ and 'typical' to be 'dense G_δ', then the italicized claim can be formalized as the following corollary.

Corollary 10.4 *Let μ be an OU measure on a compact connected manifold X. Let $\mathcal{V}$ be a conjugate invariant dense G_δ subset of $\mathcal{G}[X,\mu]$ with the weak topology. Then $\mathcal{V} \cap \mathcal{M}[X,\mu]$ is a dense G_δ subset of $\mathcal{M}[X,\mu]$, with the uniform topology.*

Proof Since the set $\mathcal{E}$ of ergodic automorphisms is also a dense G_δ set (in $\mathcal{G}[X,\mu]$ with the weak topology), so is $\mathcal{E} \cap \mathcal{V}$ by the Baire Category Theorem. Thus $\mathcal{V}$ contains an ergodic and hence antiperiodic transformation; consequently, $\mathcal{V}$ satisfies the hypotheses of the set V in Theorem 10.3, and so the conclusion follows. $\qquad\square$

This result [11] is the precise version of what we informally called Theorem C in Chapter 1. Finally we note that in [27] a theorem of this type is obtained in the closed subspace of $\mathcal{M}[X,\mu]$ with a fixed rotation vector – the definition of the rotation vector of a homeomorphism of a compact manifold is due to Fathi [58].

10.3 Applications to Fixed Point Theory

In Chapter 5 we proved (Theorem 5.5) that any orientation preserving, area preserving homeomorphism of the open square has a fixed point. We mentioned Bourgin's query [43] as to whether the restrictions to dimension 2 or to orientation preserving homeomorphisms were necessary. We can now show that these restrictions are indeed necessary to obtain existence theorems for fixed points. The Homeomorphic Measures Theorem can be used to easily construct counterexamples, that is, homeomorphisms of the cube without these properties which have no interior fixed points. The results of this section are taken from [15].

For this section we will take the n-cube to be the n-fold product of the interval $[-1, 1]$ with itself; i.e., $I^n = \{x \in R^n : -1 \le x_i \le +1\}$. Let I^n_+ and I^n_- denote the upper and lower halves of I^n determined by the conditions $x_n \ge 0$ and $x_n \le 0$, and let their common face $x_n = 0$ be denoted by F^n. Identify I^{n-1} with F^n by the embedding $p = p_n : I^{n-1} \to F^n$ given by $p(x_1, x_2, \ldots, x_{n-1}) = (x_1, x_2, \ldots, x_{n-1}, 0)$. Observe that every homeomorphism g of I^{n-1} determines a homeomorphism $\phi(g)$ of F^n by the formula $\phi(g) = pgp^{-1}$. Let α be the reflection of R^n about the hyperplane $x_n = 0$, that is $\alpha(x_1, x_2, \ldots, x_{n-1}, x_n) = (x_1, x_2, \ldots, x_{n-1}, -x_n)$. For any homeomorphism f of the boundary of I^n let $\mathcal{H}_f[I^n]$ and $\mathcal{M}_f[I^n, \mu]$ be the subsets of the unsubscripted spaces ($\mathcal{H}[I^n]$ and $\mathcal{M}[I^n, \mu]$, respectively) consisting of homeomorphisms which are equal to f on the boundary. We denote by id the identity map on the space under consideration.

Lemma 10.5 *Given any g in $\mathcal{H}_{id}[I^{n-1}]$ there is an f in $\mathcal{H}[I^n_+]$ such that $\phi(g)$ is the restriction of f to F^n and f is the identity on all the other faces of I^n_+.*

Proof Every homeomorphism of a closed ball which fixes the boundary is isotopic to the identity via an isotopy which fixes the boundary. Hence there is an isotopy $T : I^{n-1} \times [0, 1] \to I^{n-1}$ from g to the identity id. That is, T is continuous and for $0 \le s \le 1$, each map $T_s(x) = T(x, s)$ is a homeomorphism of I^{n-1}, with $T_0(x) = g(x)$, $T_1(x) = x$ and $T_s(y) = y$ for $y \in \partial I^{n-1}$. The lemma is established by defining $f(x, x_n) = (T(x, x_n), x_n)$, where $x = (x_1, x_2, \ldots, x_{n-1})$. $\square$

Lemma 10.6 *Given any homeomorphism g in $\mathcal{H}_{id}[I^{n-1}]$, there exists* *a volume preserving homeomorphism r in $\mathcal{M}_\alpha\,[I^n, \lambda]$ such that $r\left(I^n_+\right) = I^n_-$, $r\left(I^n_-\right) = I^n_+$, and r equals $\phi(g)$ on F^n.*

Proof Let f be the 'extension' of g to I^n_+ given by the previous lemma. By the Homeomorphic Measures Theorem 9.1 there is a volume preserving homeomorphism h in $\mathcal{M}\left[I^n_+, \lambda\right]$ which is the identity on all the faces of I^n_+, except for the face F^n, where it equals $\phi(g)$. Finally, define the required homeomorphism r by the rules $r(x) = \alpha h(x)$, if $x_n \geq 0$, and $r(x) = h\alpha(x)$, if $x_n \leq 0$. □

Lemma 10.7 *If there is a homeomorphism g in $\mathcal{H}_{id}\left[I^{n-1}\right]$ which has no interior fixed point, then there is an orientation reversing, volume preserving homeomorphism r in $\mathcal{M}_\alpha\,[I^n, \lambda]$ which has no interior fixed point.*

Proof Given g, let r be the homeomorphism obtained in the previous lemma. If x is an interior fixed point of r, then x must be in F^n and $p^{-1}(x)$ must be a fixed point of g, contrary to the assumption. □

Theorem 10.8 *For each $n \geq 2$, there exists an orientation reversing volume preserving homeomorphism r in $\mathcal{M}_\alpha\,[I^n, \lambda]$ which has no interior fixed point.*

Proof According to the previous lemma, we need only establish the existence of homeomorphisms of $I^n, n = 1, 2, \ldots$, which fix the boundary and have no interior fixed points. This is of course trivial. Take, for example, the homeomorphism

$$g\left(x_1, x_2, \ldots, x_{n-1}, x_n\right) = \left(x_1, x_2, \ldots, x_{n-1}, \left(\left(x_n + 1\right)^2/2\right) - 1\right).$$

□

The restriction of the above homeomorphism r to the interior of I^2 answers Bourgin's question regarding dimension 2. For orientation preserving volume preserving homeomorphisms in higher dimensions we continue as follows.

Corollary 10.9 *For $n \geq 3$ there is an orientation preserving, volume preserving homeomorphism of I^n which has no interior fixed point.*

Proof Fix any $n \geq 2$ and let r be the orientation reversing homeomorphism of I^n without interior fixed points, given by the previous theorem.   Define an orientation preserving homeomorphism g in $\mathcal{M}\left[I^{n+1}, \lambda\right]$ by $g\left(x, x_{n+1}\right) = \left(r(x), -x_{n+1}\right)$, where $x = \left(x_1, x_2, \ldots, x_{n-1}, x_n\right)$. If a point $\left(x, x_{n+1}\right)$ were an interior fixed point of g, then we would have $x_{n+1} = 0$ and hence x would be an interior fixed point of r, contrary to assumption.   □

The obstruction to extending this corollary to dimension 2 (and hence contradicting the conclusion of Theorem 5.5) is the lack of any orientation reversing homeomorphism of the interval $[-1, +1]$ which has no interior fixed point.

This corollary can also be established using differentiable techniques [36].

Part III
Measure Preserving Homeomorphisms of a
Noncompact Manifold

11

Introduction to Part III

11.1 Noncompact Manifolds

Up to now, we have considered dynamics on compact manifolds with finite measures. In this last part of the book we widen our analysis to include noncompact manifolds and consequently infinite measures.

- Topologically, the analysis extends to cover *sigma compact manifolds* X – manifolds which can be represented as a countable union of compact sets. In fact (see Section 14.6), they can be represented as a countable union of compact manifolds. As in the compact case, we allow a manifold boundary, which we denote by ∂X. For noncompact manifolds, the notion of an *end* (roughly, a way of going to infinity) will turn out to be of great importance. This notion will be introduced informally in Chapter 13, and then more formally in Chapter 14.

- Measure theoretically, the manifold X will be endowed with a fixed OU measure μ which can be finite or infinite, but in any case the definition of an OU measure ensures it is sigma finite. This means the space X can be written as a countable union of sets of finite μ-measure. Mainly we will be interested in the case where *the OU measure μ is infinite*, as the finite measure case resembles the theory developed earlier for compact manifolds. The relation between the ends of the manifold X and the measure μ will be important for the theory we will develop. Some ends will have infinite measure, and those *ends of infinite measure* will be significant in the theory.

11.2 Topologies on $\mathcal{G}[X, \mu]$ and $\mathcal{M}[X, \mu]$: Noncompact Case

As in the previous parts of the book, we are interested in the space of automorphisms (μ-preserving bijections) of (X, Σ, μ), denoted $\mathcal{G}[X, \mu]$, and particularly in the subspace $\mathcal{M}[X, \mu]$ consisting of μ-preserving homeomorphisms. Our primary focus is on the ergodic theoretic, or dynamical, properties of the homeomorphisms in $\mathcal{M}[X,\mu]$. We want to determine whether properties which are typical in $\mathcal{G}[X,\mu]$ are also typical in $\mathcal{M}[X,\mu]$, or at least in some subspace of $\mathcal{M}[X,\mu]$. To make sense of this question, we must extend our definitions of the weak and uniform topologies on $\mathcal{G}[X,\mu]$ for compact X, to the noncompact (and infinite measure) case.

In the case $\mu(X) = \infty$, the weak topology is defined as before by the sequential convergence of g_i to g in $\mathcal{G}[X,\mu]$ when $\mu(g_i B \triangle gB) \to 0$ for all measurable sets B of *finite* μ-measure. This topology is defined by the basic open neighborhoods

$$\mathcal{W}(g,\delta,B_1,\ldots,B_m) = \{f : \mu(fB_i \triangle gB_i) < \delta, i = 1,\ldots,m\},$$

where δ is a positive number and the sets B_i have finite measure. This is topologically complete. An automorphism of an infinite measure space is ergodic if its only invariant sets are those which have measure zero, or have a complement of measure zero. We can easily construct ergodic automorphisms of (X,Σ,μ) when $\mu(X) = \infty$, as follows: Given any set $A \subset X$ of finite measure, let $\hat{f}$ be any ergodic automorphism of the finite measure set A onto itself and then extend $\hat{f}$ to an ergodic automorphism f of X by viewing X as a skyscraper over A. This construction is outlined in Theorem A1.2 in Appendix 1; furthermore note that this is a generalization of the construction given earlier in the proof of Theorem 7.1 where we gave a simple proof of Theorem A1.2 that the 'skyscraper' over the ergodic automorphism $\hat{f}$ is ergodic. (See also Friedman's book [63] for a nice description of these skyscraper constructions.) More important to us is not just an *example* of an ergodic automorphism of an infinite measure space, but rather their *prevalence* in $\mathcal{G}[X,\mu]$. This important dynamical result, generalizing that of Halmos for the case of finite measure, was proved by Sachdeva [101] and Choksi and Kakutani [50].

Theorem 11.1 *Let (X,Σ,μ) be an infinite, sigma finite, Lebesgue space. The subset $\mathcal{E}$ of $\mathcal{G}[X,\mu]$, consisting of ergodic automorphisms, is dense and G_δ with respect to the weak topology.*

In addition there are many other properties which have been shown to be typical for automorphisms of an infinite Lebesgue space (X,Σ,μ) (i.e., one which is measure theoretically conjugate to the real line R^1 with length measure λ) in [101] and [50]. Typical properties include zero entropy [79], and ergodic of infinite index (all Cartesian powers are ergodic). A survey of such properties is given in Choksi and Prasad's article [51]. The book by J. Aaronson *Infinite ergodic theory* [1] has other typical properties. Finally we note the following: On an infinite measure space with no atoms every ergodic automorphism is recurrent. Thus if we have an automorphism which is not recurrent, then it cannot be ergodic.

In extending the uniform topology to the noncompact case, we have a choice. We could require uniform convergence on X, or uniform convergence on all compact subsets. We will adopt the latter course (called the compact-open topology), for the following reason. Consider the planar homeomorphism $t : R^2 \to R^2$, defined by $t(x_1, x_2) = (x_1 + 1, x_2)$, which translates to the right by one unit. Clearly t preserves planar Lebesgue measure λ, and so belongs to $\mathcal{M}[R^2, \lambda]$. Note that any homeomorphism $h \in \mathcal{M}[R^2, \lambda]$ with $|h(x) - t(x)| < 1/2$ for all $x \in R^2$ cannot be λ-recurrent. Consequently it cannot be ergodic. It follows that with respect to the *uniform* topology on $\mathcal{M}[R^2, \lambda]$, the ergodic homeomorphisms cannot be dense. Indeed, there is an open set of nonergodic homeomorphisms. Thus ergodicity is not a dense G_δ property in the space of area preserving homeomorphisms with the topology of uniform convergence. The Baire category approach to proving the existence of ergodic area preserving homeomorphisms of R^2 comes to a dead end with the topology of uniform convergence. However, if we only require for example that h is within $1/2$ of the translation t *on a large square* (i.e., in the compact-open topology), it is indeed possible for nearby h to be ergodic. By adopting the topology for which the sequence g_i converges to g if it does so uniformly *on compact sets* (the compact-open topology), we will be able to obtain genericity results for ergodic theoretic properties in $\mathcal{M}[X,\mu]$, for noncompact manifolds X. We note the analogy to the extension of the weak topology on $\mathcal{M}[X,\mu]$, when $\mu(X) = \infty$, where we defined the convergence of g_i to g in $\mathcal{G}[X,\mu]$ when $\mu(g_i B \bigtriangleup gB) \to 0$ for all measurable sets B of *finite* μ-measure. As in the compact case, we view $\mathcal{M}[X,\mu]$ as a subset of $\mathcal{G}[X,\mu]$, and define our topologies initially on the latter space. So we define the compact-open topology on $\mathcal{G}[X,\mu]$ by specifying as basic open sets the neighborhoods of $g \in \mathcal{G}[X,\mu]$ given by

$$\mathcal{C}(g, K, \delta) = \{f \in \mathcal{G}[X,\mu] : d(f(x), g(x)) < \delta, \text{ for } \mu\text{-a.e. } x \in K\}$$

where K is a compact subset of X and δ is a positive number. The relative topology on the subset $\mathcal{M}[X,\mu]$ is topologically complete. Of course when X is compact, this definition reduces to that of the uniform topology. The compact-open topology on $\mathcal{M}[X, \mu]$ can be metrized as follows: let K_i be a sequence of compact sets whose union is X. Then the compact-open topology is induced by the complete metric

$$U(f, g) = \sum_{i=1}^{\infty} \frac{u_{K_i}(f, g)}{1 + u_{K_i}(f, g)},$$

where

$$u_S(f, g) = \max \left(\max_{x \in S} |f(x) - g(x)|, \max_{x \in S} \left| f^{-1}(x) - g^{-1}(x) \right| \right).$$

This metric reduces to the earlier uniform topology on compact manifolds given in Chapter 2. We will not use this metric in most of the text except for the final chapter of this book.

11.3 Main Results for Sigma Compact Manifolds

The aim of this section is similar to that of Section 1.3, where the main results of Parts I and II were informally presented. Here we present the main results of Part III. Formal definitions of the concepts related to ends can be found in Chapter 14.

The first result says roughly that Theorem C (of Chapter 1) for compact manifolds can be extended without any additional hypotheses to Euclidean space R^n, $n \geq 2$, as long as we adopt the compact-open topology on $M[R^n, \lambda]$.

Theorem E *If a measure theoretic property is typical for automorphisms of an infinite Lebesgue measure space, it is also typical for volume preserving homeomorphisms of Euclidean space R^n, $n \geq 2$.*

It is this result, established first for the property of ergodicity by Prasad [96] and extended to arbitrary measure theoretic properties in steps by Alpern [12, 14], which initiated the study of typical dynamics on noncompact manifolds.

This positive result was followed by two counterexamples of the authors which showed that the situation for general noncompact manifolds was not the same as for Euclidean space. We shall give informal versions of these examples here (they are formally given as Examples 13.1 and 13.2).

For both examples let S denote the infinite horizontal strip manifold $R \times [0, 1]$, let $t(x, y) = (x + 1, y)$ denote the unit translation to the right, and for each integer i let $e_i = (i, 1/2)$.

For the first example, let μ be a t-invariant OU measure which is infinite on any disk centered at an e_i. Since the e_i's are invariant under the translation t, it can be viewed as a μ-preserving homeomorphism of the 'punctured strip' $S' = S - \bigcup_i e_i$. In addition to the deleted points e_i, the punctured strip has two additional 'ends', namely $e_{-\infty} = \lim_{i \to \infty} e_{-i}$ and $e_\infty = \lim_{i \to \infty} e_i$. The action induced by t on the ends is to take e_i to e_{i+1} for the finite ends, and to fix the infinite ends. Consequently t maps the end set $\{e_0, e_1, \ldots, e_\infty\}$ into its proper subset $\{e_1, \ldots, e_\infty\}$. A homeomorphism with this property is said to induce a *compressible* action on the ends and cannot be approximated by an ergodic homeomorphism in the compact open topology (see Lemma 14.15). When such a compression of ends does not arise, we say that the homeomorphism induces an *incompressible* action on the ends. An explicit example that is equivalent to the one described here is given as the Manhattan manifold in Example 13.1.

The second example of a measure preserving manifold homeomorphism which cannot be approximated by an ergodic one is much simpler: the translation $t : S \to S$ with Lebesgue (area) measure λ on S. The strip manifold S has only the two ends $e_{-\infty}$ and e_∞, and these are fixed by the action of t. Consequently the action induced by t on these two ends is incompressible. However, there is another problem with the translation t on S that prevents it from being approximated by an ergodic homeomorphism. Consider any rectangle $K = [a, b] \times [0, 1], a < b - 1$, and denote the two components of $S - K$ by $e_\infty(K) = (b, \infty) \times [0, 1]$ and $e_{-\infty}(K) = (-\infty, a) \times [0, 1]$. Observe that t has a net flow of one unit of mass to the right, which we calculate as $\mu(t(K) \cap e_\infty(K)) = 1$. We say that t induces a charge of $+1$ on e_∞ and -1 on $e_{-\infty}$. It is shown in Theorem 14.23 that a manifold homeomorphism which induces a nonzero charge on the ends cannot be approximated by an ergodic homeomorphism.

Our main result for ergodic approximation says that if the problems associated with the two above dynamical systems are excluded by hypothesis, then ergodic approximation is always possible. That is, there are no other types of problems that might arise. This result is summarized as follows (see Corollary 15.9). It has a broad range of consequences, which are developed in Chapter 15.

Theorem F *A μ-preserving homeomorphism of a sigma compact measured manifold (X, μ) is in the compact-open closure of the ergodic*

*homeomorphisms if and only if it induces an incompressible homeo-
morphism of the ends and an identically zero charge on the ends.*

Suppose now that we are interested in some more general property
$\mathcal{F}$ which is typical for automorphisms of an infinite Lebesgue space, as
in Theorem E. To approximate a homeomorphism by one with prop-
erty $\mathcal{F}$ we will need to assume that it is end preserving, that is, its
induced action on the ends is simply the identity. (An example of such
a homeomorphism is $t : S \to S$ as defined above). The following result,
formalized in Theorem 17.1, was established by the authors jointly with
Jal Choksi [18].

Theorem G *If a measure theoretic property is typical for automorphisms of an infinite Lebesgue measure space, it is also typical for end
preserving, zero charge homeomorphisms of a sigma compact manifold
which preserve a given OU measure.*

At the other extreme from fixing every end, we have established [21] a
similar result (Theorem 17.4) for homeomorphisms which behave rather
wildly on the ends. The definition of componentwise weak mixing is
given in Chapter 14.

Theorem H *If a measure theoretic property is typical for automorphisms of an infinite Lebesgue measure space, it is also typical for homeomorphisms of a sigma compact manifold which preserve a given OU
measure and whose action on the ends of the manifold is componentwise
weak mixing.*

The following version of Theorem D for noncompact manifolds can
be obtained by using Theorem G for the particular property of weak
mixing.

Theorem I *Homeomorphisms with maximal chaos are dense in the
set of end preserving, zero charge homeomorphisms of a sigma compact
manifold which preserve a given OU measure.*

11.4 Outline of Part III

We now give a brief description of the material covered in the remaining
chapters. In Chapter 12 we consider homeomorphisms of Euclidean
space R^n which preserve (n-dimensional) volume measure λ (for $n \geq 2$).
We are able to reproduce the results of Parts I and II by showing first
that ergodicity, and then that other generic properties of $\mathcal{G}[R^n, \lambda]$, are

generic in $\mathcal{M}[R^n, \lambda]$. An important property of R^n used in these proofs is that R^n is the union of an increasing family of cubes, and that these cubes have connected complements. We will see that this implies that the manifold R^n has a single end.

Chapter 13 contains the negative results that the theorems obtained for R^n cannot be extended to arbitrary noncompact manifolds. The counterexamples presented in that chapter are explained in terms of an informal discussion of the ends of a manifold. In Chapter 14 a formal definition is given of the ends of a manifold and of the two-valued (0 and ∞) measure μ^* induced on the ends by the OU measure μ. It is shown how a homeomorphism $h \in \mathcal{M}[X, \mu]$ induces a homeomorphism h^* on the ends $E[X]$ of the manifold and how h also induces a signed measure, called the *charge*, on the clopen sets of ends. We prove that if h induces a *compressible* end homeomorphism or a nonzero charge on the ends, then it cannot be μ-recurrent, or even the compact-open topology limit of recurrent homeomorphisms. These results give a general explanation of why the counterexamples of Chapter 13 work, as one induces a compressible end homeomorphism, and the other induces a nonzero charge. More importantly, we also give positive results in Chapter 15: these are consequences of the main theorem that if a homeomorphism $h \in \mathcal{M}[R^n, \mu]$ induces an incompressible end homeomorphism and a zero charge on the ends, then it is the limit of ergodic homeomorphisms. Thus we have necessary and sufficient conditions for h to be the limit of an ergodic homeomorphism. The proof of the main result needed for the positive results of Chapter 15 is presented in Chapter 16, which thus gives a complete answer for noncompact manifolds to the question of generic ergodicity in $\mathcal{M}[X, \mu]$.

In the final Chapter 17 we consider other dynamical properties, in particular those represented by a subset $\mathcal{F}$ of $\mathcal{G}[X, \mu]$ (when $\mu(X)$ is infinite) which is dense and G_δ with respect to the weak topology. One such property is weak mixing, defined as having an ergodic Cartesian square. We first settle the existence question (Is $\mathcal{F} \cap \mathcal{M}[X, \mu]$ nonempty?) by showing that if $h \in \mathcal{M}[X, \mu]$ induces the identity homeomorphism and zero charge on the ends, then it can be approximated by homeomorphisms with property $\mathcal{F}$. Since the identity homeomorphism on X has this property, there is always a μ-preserving homeomorphism of the noncompact manifold X with property $\mathcal{F}$. We conclude Chapter 17 by showing that if a homeomorphism $h \in \mathcal{M}[X, \mu]$ induces a topologically weak mixing end homeomorphism, then it can also be approximated by homeomorphisms with property $\mathcal{F}$. The results presented in Part III rely to a large extent on measure theoretic results proved in Appendix 1, and on the work of Berlanga and Epstein [38] on homeomorphic measures on noncompact manifolds, presented in Appendix 2.

12
Ergodic Volume Preserving Homeomorphisms of R^n

12.1 Introduction

The study of typical properties of volume preserving homeomorphisms of noncompact manifolds was initiated by Prasad [96], with a proof that ergodicity is generic when the manifold is Euclidean n-space R^n, $n \geq 2$. Shortly thereafter Prasad's result for ergodicity was extended by Alpern [12, 14] to all properties generic for automorphisms of an infinite Lebesgue space, but still only on the particular manifold R^n. This chapter is devoted to explaining and proving these results for R^n. Unlike the compact case, where the analysis for the cube I^n is essentially the same as for all compact manifolds, in the noncompact case the study of R^n is directly applicable only to the special class of noncompact manifolds with a single end. However, some of the ideas used here will be of general use for the noncompact setting, so this chapter will give the reader a gentle introduction to the more varied manifolds to come later.

The interest in dynamics on Euclidean space R^n dates at least to the famous *Scottish Book* [85] of 1935. This was a record kept by Polish mathematicians including Banach, Steinhaus, and Ulam, of problems discussed at the 'Scottish Cafe' in Lwów, Poland. Problem 115 of the Scottish Book, posed by Ulam, asks the following question:

Does there exist a homeomorphism h of the Euclidean space R^n with the following property? There exists a point p for which the sequence of points $h^n(p)$ is everywhere dense in the whole space.

In our terminology, this questions asks whether there is a transitive homeomorphism of R^n. This question was answered very quickly for $n = 2$, when in 1937 Besicovitch [40] gave an explicit example of a transitive homeomorphism of the plane. Since ergodicity implies transitivity (as long as open sets have positive measure), Prasad's result (Theorem

89

12.4) demonstrates that transitivity in fact represents the general case, for all dimensions $n \geq 2$. Unlike Besicovitch's explicit transitive homeomorphism of the plane, we know of no explicit ergodic homeomorphisms of any space R^n, so that Prasad's genericity result also establishes existence. For more recent work related to Ulam's problem, see [108], [109] and [84], and [16]. For other Baire category approaches to transitivity on I^n and R^n the reader is referred back to Chapter 4.

This chapter is organized as follows. In Section 12.2 we show (Lemma 12.3) that homeomorphisms with arbitrarily large invariant cubes are dense in $\mathcal{M}[R^n, \lambda]$. This fact is used in Section 12.3, in Prasad's proof of generic ergodicity (Theorem 12.4). Section 12.4 presents Alpern's proof that (Theorem 12.6) any property which is typical for automorphisms of an infinite Lebesgue space is typical in $\mathcal{M}[R^n, \lambda]$. While following the analysis for R^n given in this chapter, the reader is advised to consider which techniques would work on any noncompact manifold and which ones rely on special properties of R^n. Such considerations will motivate all the subsequent material of the book.

Before proceeding with the proofs of these results, we note that we require the dimension $n \geq 2$ because there are not many length preserving homeomorphisms of the line with interesting dynamics – in fact none of them are ergodic! If the length preserving homeomorphism is order preserving, then it must be a translation by a and so cannot be recurrent since (supposing $a > 0$) it maps the right half line into a proper subset of itself. In a sense to be made precise later on, we will say that this translation by an amount $a > 0$ has charge $+a$ to the right (or equivalently $-a$ to the left). If the length preserving homeomorphism is not order preserving then it must be a 'flip' about some point ($f(x) = 2b - x$, where b is the flip point); a flip has many nontrivial invariant sets (any symmetric interval about the flip point) and so is also not ergodic. Thus in order to have a richer supply of volume preserving homeomorphisms we restrict our attention to the case when $n \geq 2$.

12.2 Homeomorphisms of R^n with Invariant Cubes

Our first proof of Prasad's Theorem on generic ergodicity in R^n for $n \geq 2$ (Theorem 12.4) is similar to our proof in Section 7.2 of generic ergodicity in I^n. The reader is urged to review that section and in particular to read the statement of Lemma 7.2. We would like to apply Lemma 7.2 to a given volume preserving homeomorphism f of R^n. However, in general f need not have *any* invariant cubes. The purpose of this section is to show

(Lemma 12.3) that any volume preserving homeomorphism of R^n can be compact-open topology approximated by another volume preserving homeomorphism which has an arbitrarily large invariant cube. This will enable us to assume in subsequent sections that the homeomorphism we wish to approximate already has an invariant cube.

In order to construct a homeomorphism with invariant cubes, we will need a deep result in topology called the Annulus Theorem. Any space homeomorphic to the product of the $(n-1)$-sphere with the closed unit interval, $S^{n-1} \times [0,1]$, is called an *annulus*. An embedding $f : S^{n-1} \rightarrow R^n$ is said to be *locally flat* if for each $x \in S^{n-1}$, there is a neighborhood U of $f(x)$ and a homeomorphism h between $(U, U \cap f(S^{n-1}))$ and (R^n, R^{n-1}); i.e., h is a homeomorphism between U and R^n whose restriction to $U \cap f(S^{n-1})$ is a homeomorphism to R^{n-1}. The following deep result in topology was proved in parts by Freedman, Kirby and Quinn (see [78], [99]).

Theorem 12.1 Annulus Theorem (Freedman, Kirby, Quinn) *Let $f, g : S^{n-1} \rightarrow R^n$ be disjoint, locally flat, orientation preserving embeddings of the $n-1$ sphere S^{n-1}, with $f(S^{n-1})$ inside the bounded component of $R^n - g(S^{n-1})$. Then the closed region bounded by $g(S^{n-1})$ and $f(S^{n-1})$ is homeomorphic to the annulus $S^{n-1} \times [0,1]$. Furthermore, all orientation preserving self-homeomorphisms of the $(n-1)$-sphere are isotopic.*

Remark *Actually, the 'Furthermore' statement is not part of the Annulus Theorem, but was shown to be a consequence of the Annulus Theorem by Brown and Gluck [46].*

We begin the proof of Theorem 12.4 by using the Annulus Theorem and the Homeomorphic Measures Theorem for the annulus (Corollary A2.6 in Appendix 2) to show that the volume preserving homeomorphisms possessing a large invariant cube are dense in the compact-open topology in $\mathcal{M}[R^n, \lambda]$.

Lemma 12.2 *Let $f \in \mathcal{M}[R^n, \lambda]$, a compact n-cube K and a compact n-cube C containing $K \cup f(K)$ in its interior, be given. Then there is a volume preserving homeomorphism $\hat{f} \in \mathcal{M}[R^n, \lambda]$ which leaves C invariant, and agrees with f on K.*

Proof We are given $f \in \mathcal{M}[R^n, \lambda]$, K and C compact n-cubes with $K \cup f(K) \subset \text{Int } C$. We need to know that the regions $C - \text{Int } K$ and

$C - \text{Int } f(K)$ are homeomorphic. Clearly, $C - \text{Int } K$ is an annulus and the fact that $C - \text{Int } f(K)$ is an annulus follows from the Annulus Theorem because $f|\partial K$, being the restriction of an R^n homeomorphism, is a locally flat embedding into R^n. Consequently, we can assert the existence of a homeomorphism $h : C - \text{Int } K \to C - \text{Int } f(K)$. Furthermore we may use the isotopy part ('furthermore') to assume that h also agrees with f on ∂K. Observe that while h is a homeomorphism it is not necessarily volume preserving. To get a volume preserving map we define two Borel measures on the annulus $A = C - \text{Int } K$: For each Borel set $B \subset A$ let

$$\mu_1(B) = \lambda(B) \text{ and } \mu_2(B) = \lambda(hB).$$

The measures μ_1 and μ_2 are OU measures on the annulus A (nonatomic, locally positive Borel measures which are both zero on the boundary) with $\mu_1(A) = \mu_2(A)$. Therefore by the Homeomorphic Measures Theorem applied to the annulus (Corollary A2.6), there is a homeomorphism $g : A \to A$ such that $\mu_2 g(B) = \mu_1(B)$ for each Borel set $B \subset A$. We choose g so that it is the identity on the boundary of A.

Then we define $\hat{f}$ on C first by setting

$$\hat{f}(x) = \begin{cases} hg(x) & \text{for } x \in C - \text{Int } K \\ f(x) & \text{for } x \in K. \end{cases}$$

To extend $\hat{f}$ to a volume preserving homeomorphism of all of R^n we proceed in the following manner: Let $C_0 = C$ and define any sequence of cubes C_k, $k = 1, 2, \ldots$, concentric to C_0 and increasing to R^n. Extend $\hat{f}$ in any way to a homeomorphism of C_1 onto itself (e.g., by extending $\hat{f}|\partial C_0$ radially along concentric cubes). Then appealing to the Homeomorphic Measures Theorem exactly as above, we extend $\hat{f}$ to a volume preserving homeomorphism of C_1. We repeat this process for each $k = 2, 3, \ldots$, extending $\hat{f}$ from C_{k-1} to a volume preserving homeomorphism of C_k. The map $\hat{f}$ so extended to all of R^n is our required volume preserving homeomorphism of R^n which agrees with f on K and leaves the n-cube C invariant. This completes the proof of the lemma. □

It follows from the above lemma that given a volume preserving homeomorphism f and a compact-open neighborhood $\mathcal{C}(f, K, \epsilon)$ of f, there is another homeomorphism $\hat{f} \in \mathcal{C}(f, K, \epsilon)$ having an invariant cube. Furthermore, because $\hat{f}$ agrees with f on K, not only is $\hat{f} \in \mathcal{C}(f, K, \epsilon)$, but in fact $\mathcal{C}(f, K, \epsilon) = \mathcal{C}(\hat{f}, K, \epsilon)$. Thus we have established the following:

Lemma 12.3 *The volume preserving homeomorphisms of R^n having (arbitrarily large) invariant cubes are dense in $\mathcal{M}[R^n, \lambda]$ in the compact-open topology.*

12.3 Generic Ergodicity in $\mathcal{M}[R^n, \lambda]$

Given our construction of invariant cubes in the previous section, our proof of generic ergodicity is similar to that given for I^n in Section 7.2.

Theorem 12.4 *The ergodic volume preserving homeomorphisms of R^n contain a dense G_δ subset, in the compact-open topology, of the volume preserving homeomorphisms of R^n, $n \geq 2$.*

Proof Let $\hat{D}_i$, $i = 1, 2, \ldots$, be an enumeration of all dyadic subcubes of R^n, of all orders. For indices i and j corresponding to dyadic cubes of the same order, define the subset $\mathcal{F}_{ij}$ of $\mathcal{M}[R^n, \lambda]$ by

$$h \ \in \ \mathcal{F}_{ij} \text{ if for some } h\text{-invariant measurable set } A,$$

$$\lambda(A \cap \hat{D}_i) \ > \ \frac{3}{4}\lambda(\hat{D}_i) \text{ and } \lambda(A \cap \hat{D}_j) < \frac{1}{4}\lambda(\hat{D}_j).$$

For any measurable set $A \subset R^n$, if $0 < \lambda(A)$ and $0 < \lambda(R^n - A)$, then we can find dyadic sets $\hat{D}_i$ and $\hat{D}_j$ satisfying the second line in the definition of $\mathcal{F}_{ij}$ above. Consequently it follows from the definition of ergodicity that every nonergodic homeomorphism is contained in $\bigcup_{ij} \mathcal{F}_{ij}$. In the previous expression, the union is taken over pairs i, j corresponding to dyadic cubes of the same order. So by Baire's Category Theorem it remains only to show that the sets $\mathcal{F}_{ij}$ are nowhere dense, in the compact-open topology. Thus let $f \in \mathcal{F}_{ij}$, where i, j are now fixed indices. We must show that in any compact-open neighborhood $\mathcal{C}(f, K, \epsilon) \cap \mathcal{M}[R^n, \lambda]$ of f there is a volume preserving homeomorphism h which together with a whole neighborhood lies outside of $\mathcal{F}_{ij}$. We may assume that K is a compact n-cube since otherwise K could be replaced (in the basic open set $\mathcal{C}(f, K, \epsilon)$) by any large cube containing it.

Without any loss of generality, by virtue of Lemma 12.2, we can assume f leaves invariant an arbitrarily large n-cube C. We take C to be of the form $[-2^k, 2^k]^n$, large enough so that the dyadic cubes $\hat{D}_i$ and $\hat{D}_j$ (as well as $K \cup f(K)$) are all contained in the interior of C. Take a fine dyadic partition $\mathcal{D}$ of C, finer than the dyadic decomposition of C to which $\hat{D}_i$ and $\hat{D}_j$ belong. Because f is also a volume preserving homeomorphism of the n-cube C, Lemma 7.2 implies there is a volume preserving homeomorphism h of the cube C with $|h(x) - f(x)| < \epsilon$ for $x \in C$, which satisfies the conclusion of Lemma 7.2 for some closed set B and some dyadic decomposition of C finer than $\mathcal{D}$. Extend h as in Lemma 12.2 to a volume preserving homeomorphism of R^n, and note that there is some $\delta > 0$ such that any other volume preserving homeomorphism g uniformly δ-close to h on C (i.e., in $\mathcal{C}(h, C, \delta) \cap \mathcal{M}[R^n, \lambda]$) will also satisfy the conclusion of Lemma 7.2 for the same set B. Consequently, just as in the classical proof of generic ergodicity given in Chapter 7, this implies that $\mathcal{C}(h, C, \delta) \cap \mathcal{M}[R^n, \lambda]$ lies outside of $\mathcal{F}_{ij}$. □

Note that the above proof (like the classical proof of generic ergodicity given in Chapter 7) is 'intrinsic to $\mathcal{M}[R^n, \lambda]$', in the sense that the proof takes place entirely within the space of volume preserving homeomorphisms. Now consider the group $\mathcal{G}[R^n, \lambda]$ of automorphisms of R^n with the weak topology and recall the result of Choksi–Kakutani and Sachdeva (Theorem 11.1) that the ergodic automorphisms $\mathcal{E}$ form a dense G_δ set in this topological space. Because weakly open sets are open in the compact-open topology, we have our main result of this chapter.

Theorem 12.5 *The ergodic volume preserving homeomorphisms of R^n are a compact-open dense G_δ subset of $\mathcal{M}[R^n, \lambda]$.*

12.4 Other Typical Properties in $\mathcal{M}[R^n, \lambda]$

We have just shown that ergodicity is typical for volume preserving homeomorphisms of R^n, that is, in $\mathcal{M}[R^n, \lambda]$. In this section we extend that result to show that, as for compact manifolds, *any* ergodic theoretical property typical for automorphisms of the underlying measure space is also typical for volume preserving homeomorphisms. Some specific properties of this type are described in the Introduction to Part III. The results and techniques of this section may be thought of as a generalization of those of Chapter 8 to an infinite measure, noncompact setting. The main difference is the use of a stronger measure theoretic conjugacy result, Corollary A1.16 (from Appendix 1) rather than Corollary A1.12 (which was quoted in Chapter 8 as Theorem 8.4). Our theorems strengthen the conjugacy theorems for an antiperiodic infinite measure preserving automorphism due to Choksi and Kakutani [50] and Sachdeva [101]. The following extension of Prasad'sTheorem 12.5 was proved in two stages: first [12] using the Annulus Theorem (which was then open for $n = 4$), and then [14] without it. The latter proof, which we use below, benefitted from the more advanced measure theoretic result in Appendix 1, Corollary A1.16 from [13].

Theorem 12.6 *Let* $\mathcal{F}$ *be any conjugate invariant subset of* $\mathcal{G}[R^n, \lambda]$ *which is dense and* G_δ *with respect to the weak topology. Then* $\mathcal{F} \cap \mathcal{M}[R^n, \lambda]$ *is a dense* G_δ *subset of* $\mathcal{M}[R^n, \lambda]$ *with respect to the compact-open topology.*

Proof Since both the set $\mathcal{E}$ consisting of ergodic automorphisms (by Theorem 11.1) and $\mathcal{F}$ (by assumption) are dense G_δ subsets of the complete space $\mathcal{G}[R^n, \lambda]$, the Baire Category Theorem ensures that their intersection is nonempty. Hence there is an ergodic automorphism g which, together with its conjugates, belongs to the set $\mathcal{F}$.

By hypothesis, $\mathcal{F}$ is the countable intersection of open sets. Assume for the time being (until the final paragraph of this proof) that in fact $\mathcal{F}$ is itself open in the weak topology. We need to show that $\mathcal{F} \cap \mathcal{M}[R^n, \lambda]$ is dense in $\mathcal{M}[R^n, \lambda]$ with respect to the compact-open topology. This amounts to demonstrating that

$$\mathcal{F} \cap \mathcal{M}[R^n, \lambda] \cap \mathcal{C}(h, K, \epsilon) \neq \emptyset,$$

for any compact-open basic neighborhood $\mathcal{C}(h, K, \epsilon)$. To simplify the proof, we will use the denseness of ergodicity in $\mathcal{M}[R^n, \lambda]$, established in the previous section, to justify the further assumption that h is an ergodic homeomorphism. (Following this proof, we shall indicate a proof independent of that result.) We may assume that the set K is a closed n-cube.

By Corollary A1.16, applied to the ergodic automorphisms h and g and the measurable set K, there is an automorphism $\hat{g} \in \mathcal{F}$ which is conjugate to g and equal to h almost everywhere on K. In fact, since K has finite volume, we only need the finite measure case of Corollary A1.16 due to Choksi and Kakutani.

Define $\tilde{g} = h^{-1}\hat{g}$. For any $\delta > 0$, it follows that

$$\tilde{g}(K) = K, \ \|\tilde{g}\|_K < \delta, \ \text{and} \ \tilde{g} \in h^{-1}\mathcal{F}$$

where $\|\tilde{g}\|_K = \operatorname{ess\,sup}_{x \in K} |\tilde{g}(x) - x|$. In fact $\tilde{g}$ equals the identity almost everywhere on K. Since the set $h^{-1}\mathcal{F}$ is open with respect to the weak topology it follows that any automorphism in some suitable subbasic neighborhood of $\tilde{g}$ will also belong to $h^{-1}\mathcal{F}$. We may further assume that all the finite measure sets involved in the definition of this weak neighborhood, as well as their $\tilde{g}$-images, belong to some large n-cube C which contains K in its interior.

Let $\bar{g} \in \mathcal{G}\,[R^n, \lambda]$ be any automorphism which leaves C invariant and agrees with $\tilde{g}$ pointwise on K and the finite measure sets defining the weak neighborhood of $\tilde{g}$. Hence $\bar{g} \in h^{-1}\mathcal{F}$.

We now apply the Lusin Theorem (Theorem 10.2) to the automorphism $\bar{g}$, first on K where there is a norm bound of δ, and then on the annulus $C - K$, without any norm bound. This gives us a homeomorphism $h_1 \in \mathcal{M}\,[R^n, \lambda]$ which is equal to the identity on the boundary of K and on the boundary of C, and which approximates both $\tilde{g}$ and $\bar{g}$ in that

$$h_1(K) = K, \ \ \|h_1\|_K < \delta, \ \text{and} \ h_1 \in h^{-1}\mathcal{F}.$$

Since K is compact and h is continuous, it follows that for δ sufficiently small $hh_1 \in \mathcal{F} \cap \mathcal{M}\,[R^n, \lambda] \cap \mathcal{C}\,(h, K, \epsilon)$, and so this set is nonempty, as required. This shows that $\mathcal{F} \cap \mathcal{M}\,[R^n, \lambda]$ is dense in $\mathcal{M}\,[R^n, \lambda]$ with respect to the compact-open topology (under the assumption that $\mathcal{F}$ is weakly open).

Thus we have shown that the theorem is true if 'G_δ' is replaced by 'open' in both the hypothesis and conclusion. Now suppose more generally that $\mathcal{F} = \bigcap_{i=1}^{\infty} \mathcal{F}_i$, where each set $\mathcal{F}_i$ is dense and open with respect to the weak topology. By what we have shown above, each set $\mathcal{F}_i \cap \mathcal{M}\,[R^n, \lambda]$ is dense in $\mathcal{M}\,[R^n, \lambda]$ with respect to the compact-open topology. Since $\mathcal{F}_i$ is open with respect to the weak topology, it follows that $\mathcal{F}_i \cap \mathcal{M}\,[R^n, \lambda]$ is open in the compact-open topology. Consequently the Baire Category Theorem (Theorem 2.1) shows that

$$\mathcal{F} \cap \mathcal{M}\,[R^n, \lambda] = \bigcap_{i=1}^{\infty} (\mathcal{F}_i \cap \mathcal{M}\,[R^n, \lambda])$$

is a dense G_δ subset of $\mathcal{M}\,[R^n, \lambda]$ with respect to the compact-open topology. $\square$

We note that the above proof used the result, proved in the previous section, that ergodic homeomorphisms are dense in $\mathcal{M}\,[R^n, \lambda]$ with respect to the compact-open topology, when it was assumed that the neighborhood $\mathcal{C}\,(h, K, \epsilon)$ was centered at an *ergodic* homeomorphism h. With a little more work we can make the proof independent of the result on generic ergodicity. Given an arbitrary homeomorphism h at the center of a basic neighborhood, we first can compose it with a small translation, if necessary, to ensure that $\lambda\,(K \bigtriangleup hK) > 0$. We then apply

Corollary A1.13 to the modified homeomorphism. If we call this doubly modified center automorphism h again, the above proof proceeds without any changes, starting with the third paragraph.

13

Manifolds Where Ergodicity Is Not Generic

13.1 Introduction

Up to this point we have shown that ergodicity (as well as other properties) is generic for homeomorphisms of any compact manifold and for Euclidean space R^n. The reader may naturally expect that we will continue in this fashion and show that generic ergodicity holds for *any* noncompact manifold. The purpose of this chapter is to show that this is not the case by presenting two measured manifolds (X, μ) for which ergodicity is *not* generic in the space $\mathcal{M}[X, \mu]$. After presenting these two examples, we will use them to motivate the notion of an *end* of a noncompact space. We will give an informal discussion of how the behavior of a homeomorphism $h \in \mathcal{M}[X, \mu]$ with respect to the ends of the manifold X can prevent it, or any homeomorphism close to it (in the compact-open topology), from being ergodic. The two types of behavior found in the counterexamples given in this chapter (namely *compressibility* and *nonzero charge*) will have to be excluded, by hypothesis, in the following chapters. In those chapters we will give positive results on the typicality of ergodicity or other dynamical properties in certain closed subspaces of $\mathcal{M}[X, \mu]$ for general noncompact manifolds X.

13.2 Two Examples

In both of the examples of measured manifolds (X, μ) given below, the manifold X is a subset of the plane. The topology of the manifold is the relative topology it inherits from the plane, and the measure μ is simply the restriction of planar Lebesgue measure λ to the manifold. In both cases the manifold is invariant under the unit translation to the right (denoted t), and we give particular attention to this homeomorphism.

Example 13.1 *The first of the planar manifolds where ergodicity is not generic consists of a horizontal strip with an infinite number of vertical strips attached. We call this manifold* $\hat{X}$*, where*

$$\hat{X} = (R \times [0,1]) \cup \left(\bigcup_{i=-\infty}^{\infty} ([i-1/4, i+1/4] \times [1,\infty)) \right).$$

The restriction of Lebesgue measure λ *to* $\hat{X}$ *is denoted by* $\hat{\mu}$*, and the restriction to* $\hat{X}$ *of the unit translation* t *to the right,* $t(x,y) = (x+1,y)$*, is denoted by* $\hat{h}$*. For later reference we call* $\hat{X}$ *the Manhattan manifold (because of the vertical strips centered at the lines* $x = i$*, for* i *an integer).*

Example 13.2 *The second manifold is simply the horizontal strip* $\check{X} = R^1 \times [0,1]$*, with* $\check{\mu}$ *the restriction of Lebesgue measure and* $\check{h}$ *the restriction of the unit translation* $t(x,y) = (x+1,y)$*. We will refer to* $(\check{X}, \check{\mu}, \check{h})$ *as the right unit translation on the strip.*

Of course it is clear in both cases (manifolds $\hat{X}$ and $\check{X}$) that the translation t is not recurrent, and hence not ergodic. To see this observe that no point in the set $C = [3/8, 5/8] \times [0,1]$ ever returns to C under the homeomorphism t. To prove that these manifolds are counterexamples to generic ergodicity however, we have to show moreover that any area preserving homeomorphism *near enough* to t (in the compact-open topology) is also nonrecurrent. To prove nonrecurrence we will use the following simple property of recurrent automorphisms.

Lemma 13.3 *Let* $g \in \mathcal{G}[X, \mu]$ *be a* μ*-recurrent automorphism of the measure space* (X, μ)*. Then for any measurable set* V *we have*

$$\mu(g(V) - V) = \mu(V - g(V)). \tag{13.1}$$

In fact this property characterizes μ*-recurrence.*

Proof We may assume without loss of generality that $\mu(V) > 0$ and that V is not g-invariant, as otherwise the equation is satisfied as $0 = 0$. It follows that at least one of the sets $A = g(V) - V$ and $B = V - g(V)$ has positive measure, and consequently also their (disjoint) union $Y = A \cup B$. Since g is μ-recurrent, the *first return map* $\tau : Y \to Y$ (given by $\tau(x) = g^m(x)$, where m is the least positive integer with $g^m(x)$ in Y; see Appendix 1) is a well defined μ-preserving automorphism of Y. Observe that for any $x \in B$, the g-orbit of x cannot re-enter B before entering A.

Hence $\tau(B) \subset A$, and consequently $\mu(B) \leq \mu(A)$. The reverse inequality is established by replacing g with its inverse, which is also μ-recurrent. So equation (13.1) holds for μ-recurrent automorphisms.

To show that (13.1) characterizes μ-recurrence, suppose now that g is not μ-recurrent. Then there is a measurable set C of positive μ-measure with the property that C and all its forward iterates $g^i(C)$, $i > 0$, are disjoint. Then the set

$$V = \bigcup_{i=0}^{\infty} g^i(C)$$

is a set such that (13.1) does not hold (i.e., $\mu(g(V) - V) = 0 \neq \mu(V - g(V))$). □

Although the difference of the two sides in equation (13.1) need not be defined for all sets V (both sides could be infinite), the following lemma shows that the difference makes sense and is zero for all automorphisms of (X, μ) whenever $\mu(V) < \infty$. Consequently, in a finite measure space ($\mu(X) < \infty$), every automorphism is μ-recurrent (this is the Poincaré Recurrence Theorem 5.4) and the difference of the two sides in equation (13.1) will be zero for all $V \subset X$.

Lemma 13.4 *Let $g \in \mathcal{G}[X, \mu]$ be any μ-preserving automorphism (not necessarily μ-recurrent) and let V be any measurable set of finite μ-measure. It follows that*

$$\mu(V - g(V)) - \mu(g(V) - V) = 0. \tag{13.2}$$

Proof Observe that

$$V = (V - g(V)) \cup (V \cap g(V))$$

and

$$g(V) = (g(V) - V) \cup (V \cap g(V)).$$

Since $\mu(V) = \mu(g(V))$ and all the sets above all have finite measure, it follows that $\mu(V - g(V)) - \mu(g(V) - V) = 0$. □

We can now use Lemma 13.3 to show that recurrence and hence ergodicity is not generic, or even dense, in the spaces $\mathcal{M}[\hat{X}, \hat{\mu}]$ and $\mathcal{M}[\check{X}, \check{\mu}]$. There are two qualitatively distinct ways in which the recurrence equation (13.1) can fail: one side can be finite when the other is infinite, or the two sides can be finite but unequal. The first type of failure applies to the right unit translation $\hat{h} \in \mathcal{M}[\hat{X}, \hat{\mu}]$ of the Manhattan manifold, and the second to the translation $\check{h} \in \mathcal{M}[\check{X}, \check{\mu}]$ of the strip manifold.

Theorem 13.5 *There is a compact-open neighborhood of the right unit translation $\hat{h} \in \mathcal{M}[\hat{X}, \hat{\mu}]$ of the Manhattan manifold consisting of nonrecurrent (and hence nonergodic) homeomorphisms. Consequently ergodicity is not a typical property in the space $\mathcal{M}[\hat{X}, \hat{\mu}]$.*

Proof Define $C = [3/8, 5/8] \times [0, 1]$, and observe that $\hat{X} - C$ has two components, both unbounded, namely $U = \hat{X} \cap \{(x, y) : x < 3/8\}$ and $V = \hat{X} \cap \{(x, y) : x > 5/8\}$. (We will later call such a set C a *separating set*.) Any $h \in \mathcal{M}[\hat{X}, \hat{\mu}]$ which is sufficiently close to $\hat{h}$ on the set C will satisfy $\mu(h(V) - V) < \infty$ (in fact $h(V) \subset V$ so $\mu(h(V) - V) = 0$) and $\mu(V - h(V)) = \infty$, so the recurrence equation (13.1) in Lemma 13.3 fails dramatically and consequently h cannot be recurrent. □

A similar argument applies to $\check{h} \in \mathcal{M}[\check{X}, \check{\mu}]$, as follows.

Theorem 13.6 *There is a compact-open neighborhood of the right unit translation $\check{h} \in \mathcal{M}[\check{X}, \check{\mu}]$ consisting of nonrecurrent (and hence nonergodic) homeomorphisms of the strip manifold. Consequently ergodicity is not generic in the space $\mathcal{M}[\check{X}, \check{\mu}]$.*

Proof Let C and V be defined as in the previous proof, but with respect to $\check{X}$, that is, $V = \check{X} \cap \{(x, y) : x > 5/8\}$. Observe that for any $h \in \mathcal{M}[\check{X}, \check{\mu}]$ which is sufficiently close to $\check{h}$ on C, we will have that $\mu(h(V) - V) = 0$ and $\mu(V - h(V)) > 0$, so that the recurrence equation (13.1) is not satisfied. Hence by Lemma 13.3 such homeomorphisms h cannot be μ-recurrent and therefore cannot be ergodic. □

The above proof is a bit too special to be generalized to other measured manifolds (X, μ). A more general method is to observe that

$$\mu(V - \check{h}(V)) - \mu(\check{h}(V) - V) > 0,$$

and that the expression

$$\mu(V - h(V)) - \mu(h(V) - V) \tag{13.3}$$

is continuous in h, in the compact-open topology. Consequently (13.3) will be positive in some compact-open neighborhood of $\check{h}$, and so Theorem 13.6 follows from Lemma 13.3. The expression (13.3) will later be called the 'charge' and is a measure of the net 'flow of mass into V'

(under the action of h) or into the end at $+\infty$. Similar ideas will show that homeomorphisms with nonzero charge cannot be approximated by ergodic ones.

13.3 Ends of a Manifold: Informal Introduction

In order to prove positive results in the remainder of the book regarding generic ergodicity (and other properties), we must exclude the behavior exhibited by the two examples in the previous section. To do so, we must have a more general framework in which to describe this behavior. This general framework involves the following notions:

(i) The *ends* of a noncompact manifold
(ii) The *induced homeomorphism* on the ends
(iii) The *compressibility* of the induced homeomorphism
(iv) The *charge* on an invariant set of ends.

In this section we give a nonrigorous, informal introduction to these notions, by seeing how they apply to the two dynamical systems, $(\hat{X}, \hat{\mu}, \hat{h})$ and $(\check{X}, \check{\mu}, \check{h})$, which we have just defined. In the next chapter we shall give formal definitions of these notions and rigorous proofs of their properties.

An end of a manifold is, informally, a way of going to infinity on that manifold. For the strip $\check{X}$ there are two ways of going to infinity: left or right. Consequently the set $\check{E}$ of ends of $\check{X}$ may be written as $\check{E} = \{-\infty, +\infty\}$. Similarly on the manifold $\hat{X}$ we may go to infinity by going left or right, giving ends $-\infty, +\infty$. But we may also go up from the x-axis at the integer values $x = i$, giving additional ends at each integer i. Thus the set of all ends of $\hat{X}$ may be written as $\hat{E} = \{-\infty, \ldots, -1, 0, 1, \ldots, +\infty\}$. The set of ends E of a manifold X inherits a topology from the manifold X (such that $X \cup E$ is compact). The topology on $\check{E}$ is simply the discrete topology and the topology on $\hat{E}$ has the usual notion of a sequence of integers converging to $+\infty$ or $-\infty$ (a basic open set in $\hat{E}$ is an interval in the extended integers $[i, j]$ where $i < j$ are extended integers that could be $-\infty$ or $+\infty$).

Every homeomorphism h of a manifold induces a homeomorphism h^* on its ends. The induced homeomorphism $\check{h}^* : \check{E} \to \check{E}$ is simply the identity, and the induced homeomorphism $\hat{h}^* : \hat{E} \to \hat{E}$ fixes $\pm\infty$ and maps each integer i to $i + 1$.

Observe that the induced homeomorphism $\hat{h}^* : \hat{E} \to \hat{E}$ maps the clopen (closed and open) set $\{1, 2, \ldots, +\infty\}$ into the proper clopen subset $\{2, 3, \ldots, +\infty\}$. Such an induced homeomorphism which maps a set into a proper subset is said to be *compressible* (see Definition 14.11). A generalization of Theorem 13.5 of this chapter, Lemma 14.15, shows that any homeomorphism $h \in \mathcal{M}[X, \mu]$ which induces a compressible homeomorphism on the ends of X cannot be recurrent, and hence cannot be ergodic. In Chapter 15 a number of positive results regarding generic ergodicity are proved under assumptions concerning incompressibility of induced end homeomorphisms.

The nonrecurrence of all homeomorphisms in a neighborhood of the dynamical system $(\check{X}, \check{\mu}, \check{h})$ cannot be explained in terms of compressibility on ends, because the induced end homeomorphism $\check{h}^*$ is simply the identity, which is incompressible (see Definition 14.11). It has to be explained in terms of the *charge* it induces on the invariant sets of ends, namely $\{+\infty\}$ and $\{-\infty\}$. The charge c_h that an end preserving homeomorphism $h \in \mathcal{M}\left[\check{X}, \check{\mu}\right]$ induces on the end called $+\infty$ is defined by the difference given by equation (13.3), namely

$$c_h\left(+\infty\right) = \mu\left(V - h(V)\right) - \mu\left(h(V) - V\right) \tag{13.4}$$

where V is as in the proof of Theorem 13.6, namely $V = \check{X} \cap \{(x, y) : x > 5/8\}$. In particular $c_{\check{h}}(+\infty) = 1$. The argument given above after the proof that $\check{h}$ is nonrecurrent (which constitutes an alternative proof) will be generalized in Theorem 14.23 to show that any homeomorphism $h \in \mathcal{M}[X, \mu]$ which induces a nonzero charge on an invariant set of ends cannot be the limit of ergodic homeomorphisms.

Both of the counterexamples to generic ergodicity given in this chapter can be excluded by restricting to those homeomorphisms in $\mathcal{M}[X, \mu]$ which induce an incompressible homeomorphism on the ends of X, and an identically zero charge on the invariant set of ends. For this closed subspace of $\mathcal{M}[X, \mu]$, ergodicity is indeed generic. Summarizing the positive results to follow in Chapter 15, we have

Theorem F (see Corollary 15.9) *A μ-preserving homeomorphism h of (X, μ) is in the compact-open closure of the ergodic homeomorphisms if and only if h induces an incompressible homeomorphism of the ends and induces an identically zero c_h charge on the ends.*

We conclude this chapter with an informal look at four examples in the following sections. Formal definitions are given in the next chapter.

13.4 Another Look at R^n

We now take a short look backwards at the results of the previous chapter concerning generic ergodicity in $\mathcal{M}[R^n, \lambda]$. Euclidean space R^n (for $n \geq 2$) has a single end, so that for any $h \in \mathcal{M}[R^n, \lambda]$, the induced end homeomorphism h^* is simply the identity on this end, and hence incompressible. Since the charges on all the invariant sets of ends sum to zero (this will be shown formally later), the charge is always identically zero when there is a single invariant set of ends. Thus in particular, when the manifold has only one end, the charge is identically zero. Hence Theorem F says that every homeomorphism in $\mathcal{M}[R^n, \lambda]$ is the compact-open limit of ergodic ones.

13.5 The Flip on the Strip

The analysis given above for R^n is similar to that for the flip homeomorphism $\check{f} : \check{X} \to \check{X}$ defined by $\check{f}(x,y) = (-x, y)$, where $\check{X}$ is the infinite horizontal strip $R^1 \times [0, 1]$. The induced end homeomorphism $\check{f}^*$ transposes the two ends $+\infty$ and $-\infty$, so it is incompressible. Since there is only one nonempty invariant set of ends, namely the full set, the induced charge is identically zero. The flip homeomorphism $\check{f}$ is clearly recurrent (all points have period 2) and not ergodic (horizontal strips are invariant). However, since it induces an incompressible end homeomorphism and zero charge, Theorem F shows that it is approximable by ergodic homeomorphisms, in the compact-open topology.

13.6 The Flip on Manhattan

Consider the flip homeomorphism $\hat{f} : \hat{X} \to \hat{X}$ defined on the Manhattan manifold $\hat{X}$ by $\hat{f}(x,y) = (-x, y)$. As in the previous example, $\hat{f}$ is recurrent but not ergodic. The induced homeomorphism $\hat{f}^*$ on the ends $E = \{-\infty, \ldots, -3, -2, -1, 0, 1, \ldots, +\infty\}$ is given by $\hat{f}^*(-\infty) = +\infty$, $\hat{f}^*(+\infty) = -\infty$, and $\hat{f}^*(i) = -i$ for integers i. Clearly $\hat{f}^*$ is incompressible. The 'basic' invariant end sets are $\{0\}, \{-1, 1\}, \{-2, 2\}, \ldots, \{-\infty, +\infty\}$ and the induced charge is zero on all of these. Hence Theorem F shows that $\hat{f}$ is the compact-open limit of ergodic area preserving homeomorphisms.

13.7 Shear Map on the Strip

The previous three examples were recurrent (even periodic) and nonergodic. Consider the shear map $s(x, y) = (x + y, y)$ on the extended strip $R \times [-1, 1]$. Observe that s^* is the identity on the end set $\{-\infty, +\infty\}$ (and hence incompressible) and has zero charge (zero net flow to the right). It is not recurrent (take any set above the x-axis) and hence nonergodic. But Theorem F says that it is the compact-open limit of ergodic area preserving homeomorphisms.

14

Noncompact Manifolds and Ends

14.1 Introduction

The previous chapter showed that the notion of the ends of a manifold provides an important distinction between Euclidean space (R^n, λ), where ergodicity is typical for volume preserving homeomorphisms, and the strip $(R^1 \times [0,1], \lambda)$, where ergodicity is not typical for volume preserving homeomorphisms. The distinction between the spaces is that R^n (for $n \geq 2$) has only one end, whereas the strip has two. In this chapter we present a formal description of the ends of the manifold X, and related topological results. This will enable us to obtain in the following chapter conditions on the ends of a measured manifold (X, μ) under which ergodicity is typical in certain closed subspaces of $\mathcal{M}[X, \mu]$.

We now provide the formal definitions and results mentioned in this informal discussion. Most of these results come from [20] and [22].

14.2 End Compactification

The notion of an end of a manifold has already been discussed informally. We now give a formal treatment.

Definition 14.1 *An* end *e of the manifold X is a function which assigns to each compact subset K of X a nonempty unbounded component $e(K)$ of $X - K$ in such a way that*

$$K_1 \subset K_2 \ \text{implies} \ e(K_2) \subset e(K_1). \tag{14.1}$$

The set of all ends of X is denoted $E[X]$.

Note that if X is itself compact then it has no unbounded subsets. The notion of an end only has significance for us when the manifold X is not compact.

Observe that because X is a manifold, $X - K$ has only a finite number of unbounded components (see Lemma A2.10 from Appendix 2) and for each end e only one of those components of $X - K$ (namely $e(K)$) 'leads to' e. By adjoining to a compact set K the union of all the bounded components of $X - K$, we obtain a larger compact set $\hat{K}$ whose complement has no unbounded components. Furthermore the components of $X - \hat{K}$ are the same as the unbounded components of $X - K$.

The manifold X is compactified by adjoining the set of ends $E[X]$, and defining for each compact set $K \subset X$ a basic neighborhood $N_K(e_0)$ of an end $e_0 \in E[X]$, as the set

$$ N_K(e_0) = e_0(K) \cup \{e \in E[X] : e(K) = e_0(K)\}. $$

With this topology, $X \cup E[X]$ is a compact Hausdorff space containing $E[X]$ as a closed subset. Again because for each compact set $K \subset X$ there are only finitely many unbounded components of $X - K$, these neighborhoods $N_K(e_0) \cap E[X]$ form a basis of closed and open sets for e_0 in $E[X]$; thus, with the relative topology on $E[X]$, the ends form a totally disconnected set.

14.3 Examples of End Compactifications

We now reconsider some of the examples given informally in the previous chapters. For $n \geq 2$, we have noted that R^n has a single end. The 'end compactification' topology of $R^n \cup E[R^n]$ given by the above definition is the usual one-point compactification which makes it into a topological sphere (homeomorphic to S^n). The cylinder $R^1 \times S^1$ however has two ends since compact sets such as $[a, b] \times S^1$ divide the space into two unbounded components, one of them leading to point at '$-\infty$' on the cylinder (the left end) and the other leading to the point at '$+\infty$' (the right end). Here again the compactification of the cylinder is the sphere S^2 and here the end set $E \subset S^2$ consists of two points. In general one can start with a compact manifold like S^2 and let $E \subset S^2$ be any totally disconnected set. Then E is the end set for the noncompact manifold $S^2 - E$. Although we will not need to deal with such problems we do note that the end sets may be embedded in S^2 in a very complicated fashion. For example when we take E to be the wild Cantor set as in Alexander's horned sphere, then the components $e(K)$ are intertwined

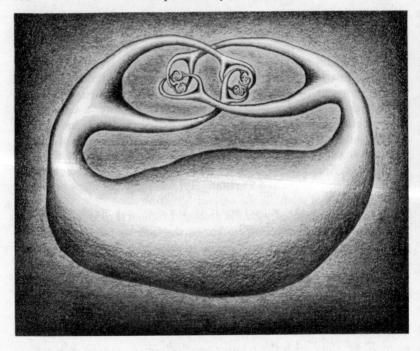

Fig. 14.1. Alexander's horned sphere (with permission from [73])

in an intricate manner. A picture of Alexander's horned sphere from Hocking and Young's *Topology* textbook [73, p. 176] is reproduced here in Figure 14.1.

14.4 Algebra $\mathcal{Q}$ of Clopen Sets

We return to the general setting of a sigma compact manifold X with end set $E[X]$. For any subset $Q \subset E[X]$, and compact set $K \subset X$, we define

$$Q(K) = \bigcup_{e \in Q} e(K).$$

Each compact set $K \subset X$ determines the following equivalence relation $\sim_K$ on $E[X]$, namely all the ends 'contained in' the component $e(K)$:

$$e \sim_K e' \text{ if and only if } e(K) = e'(K). \tag{14.2}$$

For each compact set $K \subset X$ there are only finitely many unbounded components in $X - K$. Let $\mathcal{P}_K$ denote the finite partition of $E[X]$ into equivalence classes modulo $\sim_K$ and let $\mathcal{Q}_K$ denote the finite algebra generated by $\mathcal{P}_K$. That is, the elements of $\mathcal{P}_K$ are the atoms of the algebra $\mathcal{Q}_K$. Observe that the algebra $\mathcal{Q} = \bigcup_K \mathcal{Q}_K$ (where the union is taken as K ranges through the family of compact subsets of X) is identical with the family of closed-open (clopen) subsets of $E[X]$, in the relative topology on $E[X]$, of the end compactification of X. It is on this algebra that we will define our measure induced on the ends $E[X]$. This analysis shows more clearly that $E[X]$ is totally disconnected.

14.5 Measures on Ends

An OU measure μ on X induces a two-valued (0 and ∞) measure μ^* on $(E[X], \mathcal{Q})$ as follows. If $Q \in \mathcal{Q}$, then Q belongs to $\mathcal{Q}_K$ for some compact set K. Define $\mu^*(Q) = 0$ if $\mu(Q(K)) < \infty$ and $\mu^*(Q) = \infty$ if $\mu(Q(K)) = \infty$.

Remark First we note that this definition of μ^* is independent of the compact set K once K is large enough so that $Q \in \mathcal{Q}_K$. So suppose Q is also in $\mathcal{Q}_{K'}$ for some compact set K' and for now we first make the special assumption that $K \subset K'$. *We need to show that* $\mu(Q(K)) = \infty$ *if and only if* $\mu(Q(K')) = \infty$ (the general situation will then follow from this special case). Now note that $Q(K)$ is the finite union of connected components of $X - K$ and $Q(K) = Q(K') \cup (K' \cap Q(K))$. Thus since the OU measure μ is finite on the compact set $K' \cap Q(K)$, it follows that either both $Q(K)$ and $Q(K')$ are finite μ-measured sets or they are both infinite μ-measured sets. For two arbitrary compact sets K_1 and K_2, if $Q \in \mathcal{Q}_{K_1} \cap \mathcal{Q}_{K_2}$, then let $K' = K_1 \cup K_2$. Then by the above argument $Q(K_1)$ has infinite measure if and only if $Q(K')$ does and has infinite measure if and only if $Q(K_2)$ does. This shows that the set function μ^* is well defined.

Lemma 14.2 *The set function* μ^* *taking the two values 0 and* ∞ *is a measure on the algebra of clopen sets* $\mathcal{Q}$ *on* $E[X]$.

Proof First we note μ^* is trivially a finitely additive set function. Indeed let $Q_1, Q_2, \ldots, Q_k \in \mathcal{Q}$. Let K be a large enough compact set so that $Q_1, Q_2, \ldots, Q_k \in \mathcal{Q}_K$. Then $\bigcup_{i=1}^{k} Q_i(K)$ has infinite μ-measure if and only if one of the $Q_i(K)$ does. Thus $\mu^*(\bigcup_{i=1}^{k} Q_i) = \sum_{i=1}^{k} \mu^*(Q_i)$. The proof is completed by noting that the algebra of clopen sets on the totally disconnected set $E[X]$ has the property that any countable union of disjoint clopen sets can be written as a finite union of these clopen sets.   □

Thus μ^* is a measure on $(E[X], \mathcal{Q})$ taking on only the two values, 0 and ∞. The measure is nontrivial $(\mu^*(E[X]) \neq 0)$ as long as $\mu(X) = \infty$. We again remind the reader that the measurable sets (the clopen sets) $\mathcal{Q}$ constitute only an algebra, and *not* a sigma algebra. Finally we say that the end $e \in E[X]$ is an end of infinite measure if and only if $\mu(e(K)) = \infty$ for all compact sets K. Let $E^\infty[X]$ denote the set of *ends of infinite measure*.

Lemma 14.3 *The set of infinite measured ends $E^\infty[X]$ is a closed subset of $E[X]$.*

Proof Suppose that the end e_0 is a limit point of infinite measured ends. For any compact set $K \subset X$, $\mu(e_0(K)) = \infty$ because the neighborhood of e_0, $N_K(e_0)$, contains an infinite measured end $e \in E^\infty[X]$ and $e(K) = e_0(K)$. Since $\mu(e(K))$ is infinite, so too is $\mu(e_0(K))$.   □

Note that the set $E^\infty[X]$ may not be in the algebra $\mathcal{Q}$ of clopen sets. An example may be obtained on the Manhattan manifold $\hat{X}$ of Example 13.1. Take as the measure μ a measure equal to area measure for $y \leq 1$ and such that μ is finite on each of the vertical strips. In this case $E^\infty[\hat{X}]$ consists only of the 'horizontal' ends called $+\infty$ and $-\infty$, which are limits of the finite measured ends $\{\ldots, -1, 0, 1, \ldots\}$. So $E^\infty[\hat{X}]$ is not open and hence not clopen.

We end this section by giving some more examples of noncompact manifolds X with OU measures μ and the measures μ^* on the ends $E[X]$.

Example 14.4 (See [67]) *Let $X = R^2 - \{\bar{0}\}$. Then this manifold has two ends, one of them identifiable with the origin $\bar{0} = (0,0)$ and the other with 'the point at infinity' in R^2 – this is easily seen by taking K to be a Jordan curve with the origin inside. Consider the following two different OU measures obtained by integrating the following 2-forms*

$$\omega = dx \times dy, \quad and \quad \tau = (dx \times dy)/F(x^2 + y^2)$$

where $F : (0, \infty) \rightarrow (0, \infty)$ is any smooth function with the property

$$F(r) = 1 \quad if \quad r \geq 1 \quad and \quad F(r) = r^2 \quad if \quad 0 < r < 1/2.$$

This means (abusing notation) that $\tau(A) = \int_A \tau$ and similarly $\omega(A) =$

$\int_A \omega$. *Note that ω is just our area measure on the plane. The point at infinity has infinite measure for each of these measures but (the end at) the origin is infinite for only one of these measures: $\omega^*(\bar{0}) = 0$, $\tau^*(\bar{0}) = \infty$.*

Other noncompact manifolds and measures may be obtained similarly. Let $G(x,y)$ be a smooth function on $R^2 - E$ which is positive at all points except at some set E of 'singular points for G', where by a singularity at $p \in E$ we mean that $G(p_k) \to \infty$ for any sequence of points p_k in $R^2 - E$, converging to the point $p \in E$. The singularities will correspond to the ends of the manifold. We may define an OU measure ν_G on $R^2 - E$ by the equation $\nu_G(A) = \int_A G(x,y)dxdy$, for any Borel set A. By choosing the behavior at the 'singularities' of G appropriately (i.e., so that the improper integral around a singular point is finite or infinite), we can obtain ends of finite or infinite measure.

Example 14.5 *A noncompact manifold can be obtained from a compact manifold D by removing some closed totally disconnected subset C from it. So consider the manifold $X = D - C$ where $D = \{(x,y) : x^2 + y^2 \le 1\}$ is the closed unit disk in the plane and C is the standard Cantor ternary set lying on the line $[-1/2, 1/2]$ along the x-axis. The Cantor set C may be identified with the set $E[X]$ of ends of X. We now give an example of an OU measure on X. Let $C(0)$ and $C(1)$ denote the left and right thirds of the interval $[-1/2, 1/2]$ on the x-axis. For $i_k = 0, 1$ and $m \ge 1$, let $C(i_1, \ldots, i_m, 0)$ and $C(i_1, \ldots, i_m, 1)$ be the left and right thirds of $C(i_1, \ldots, i_m)$ respectively. Let λ_1 and λ_2 denote respectively 1- and 2-dimensional Lebesgue measure (i.e., length and area measures respectively). For each Borel subset A of X define $\mu(A)$ by the formula*

$$\mu(A) = \lambda_2(A) + \sum_{m=1}^{\infty} \sum_{i_1,\ldots,i_m} 3^m \lambda_1(A \cap C(i_1, \ldots, i_m)).$$

Then note that all of the ends of X (namely the points in the Cantor set C) have infinite μ^*-measure.

Another OU measure on the same manifold $X = D - C$ which gives all ends infinite measure is the following. Let $R(i_1, \ldots, i_m)$ be the closed rectangular $3^{-(m+1)}$-neighborhood of $C(i_1, \ldots, i_m)$ in D and let

$$K_m = D - \bigcup_{i_1,\ldots,i_m} \text{Int } R(i_1, \ldots, i_m).$$

Then the sets $L_m = K_{m+1} - K_m$ consist of 2^m congruent components each with volume

$$a_m = \lambda_2(L_m)/2^m.$$

For each Borel subset A of X define $\nu(A)$ by the formula

$$\nu(A) = \lambda_2(A) + \sum_m (1/a_m)\lambda_2(A \cap L_m).$$

Then ν is an infinite OU measure on $D - C$ which gives every end in C infinite measure.

Finally we note that there is nothing special about dimension 2 in any of these examples. We could just as easily have done all of this in dimensions $n > 2$.

14.6 Compact Separating Sets

Results from topology show that a general sigma compact manifold X can be written as an increasing sequence of compact connected manifolds with special properties. We describe these compact connected submanifolds and these special properties. The proofs of some of these results may be found in our Appendix 2.

We say that $K \subset X$ is an *n-cell* if it is homeomorphic to the closed unit n-cube I^n. A set $K \subset X$ is called a *relative n-cell* if there exists a continuous function $\phi : I^n \to K$ such that

(i) ϕ is onto
(ii) ϕ restricted to Int I^n is a homeomorphism onto its image
(iii) $\phi^{-1}\phi(\partial I^n) = \partial I^n$.

In other words, a relative n-cell is a subset of X which can be obtained from the n-cube by making boundary identifications, namely a compact connected n-manifold. Recall that in Chapter 9 we stated Brown's Theorem (Theorem 9.6) which said that every compact connected n-manifold is a relative n-cell.

An important fact about a sigma compact connected n-manifold is that it can be written as the countable union of an increasing family of compact connected manifolds. In [20], it is shown how this fact follows from the deep results of Kirby, Siebenmann and Quinn ([78] and [99]). However, the weaker result that the manifold is the increasing union of relative n-cells follows from the work of Berlanga and Epstein [38] (see Lemma A2.9 in Appendix 2). Furthermore, let μ be an OU measure on X (i.e., μ is a sigma finite nonatomic Borel measure which is positive on open sets, and zero on the boundary of X – we note that the fact that μ is a Borel measure implies that μ is also finite on compact sets). When the manifold X has nonempty boundary ∂X and K is a subset of X, we denote by Bdry K the union of the topological frontier of K and $\partial X \cap K$. Not only can we choose a sequence of relative n-cells increasing to X, but it follows from Berlanga and Epstein's work that the relative n-cells can be chosen so that their boundaries have measure zero. This will follow from the theorem below.

Theorem 14.6 *Let μ be an OU measure on a sigma compact connected n-manifold X. Then any compact subset C of X is contained in the interior of a relative n-cell K such that $X - K$ has no bounded components and $\mu(\text{Bdry } K) = 0$.*

Proof First apply Lemma A2.9 with $A = \emptyset$ and $B = C$ the given compact set. Then since $X - A$ is connected (the 'furthermore' part of the Lemma states) there is a single relative n-cell which we call L_1 containing C. Letting $\hat{L}_1$ be the union of L_1 and all of the bounded (compact) components of $X - L_1$, Lemma A2.10 implies that $\hat{L}_1$ is a compact set with only unbounded components in its complement. Another application of Lemma A2.9 but this time with $A = \emptyset$ and $B = \hat{L}_1$ gives the required relative n-cell K. □

Definition 14.7 *A relative n-cell K such that $X - K$ has only unbounded components, and $\mu(\text{Bdry } K) = 0$, is called a separating set.*

Lemma 14.8 *Let μ be an OU measure on a sigma compact connected manifold X, and μ^* the measure induced on $(E[X], \mathcal{Q})$. Let $Q \in \mathcal{Q}$ be a clopen set of ends in $E[X]$. Then there is a separating set $K \subset X$ such that $Q \in \mathcal{Q}_K$.*

Proof Since $Q \in \mathcal{Q}$, there is a compact set K_1 such that $Q \in \mathcal{Q}_{K_1}$. By the above theorem, there is a separating set K containing K_1 in its interior. □

14.7 End Preserving Lusin Theorem

In the later portions of the book, when proving genericity results for homeomorphisms of noncompact manifolds, we will need an extension of our Lusin Theorem 10.2 which preserves the end structure of the manifold. This extension is given below as Theorem 14.9. To motivate this extension, we refer briefly to the proof of Theorem 12.6, which gave a genericity result related to the manifold R^n. In that proof we used the Lusin Theorem 10.2 (actually, we used it twice) to approximate an automorphism $\bar{g}$ which left an n-cube K invariant and had small norm on K (that is, $|\bar{g}(x) - x| < \epsilon$ for λ-a.e. x in K). The approximating homeomorphism $h_1 \in \mathcal{M}[R^n, \lambda]$ obtained via Theorem 10.2 also had small norm on K, left K invariant, and *consequently* left $e_\infty(K) = R^n - K$ invariant. Here we use the notation e_∞ to denote the single end at infinity of R^n. The invariance of $e_\infty(K)$ is a trivial consequence of the fact that there is a single end, since whenever a set is invariant under a bijection, so is its complement. However, when there is more than one end, the invariance of K does not ensure the invariance of the sets $e(K)$, so this has to be part of the conclusion of any Lusin theorem that we use. Of course, there are only finitely many sets of the form $e(K)$. These are the sets $P(K)$, for $P \in \mathcal{P}_K$. So we give the needed Lusin theorem in the following form.

Theorem 14.9 (End Preserving Lusin Theorem) *Let μ be an OU measure on the sigma compact metric manifold (X, d) and let K be any separating subset of X. Let $g \in \mathcal{G}[X, \mu]$ be any automorphism of X satisfying*

(i) $d(g(x), x) < \epsilon$ *for μ-a.e. $x \in K$*
(ii) $g(K) = K$
(iii) $g(P(K)) = P(K)$ *for every $P \in \mathcal{P}_K$.*

Then any weak topology neighborhood of g contains a homeomorphism $h \in \mathcal{M}[X, \mu]$ with compact support which also satisfies properties (i)–(iii) (with h replacing g). We note that condition (iii) is equivalent to the condition

(iii′) $g(Q(K)) = Q(K)$ *for every $Q \in \mathcal{Q}_K$.*

Proof The proof is similar to the part of the proof of Theorem 12.6 (generic properties for R^n) where the compact form of the Lusin Theorem 10.2 was used twice: on the n-cube K with a norm bound, and on the annulus $C - K$ without a norm bound. Here we will use Theorem 10.2 $j + 1$ times, where j is the cardinality of $\mathcal{P}_K$: once on the separating set K of the theorem, and once for each component $P(K)$ where $P \in \mathcal{P}_K$.

The application of Theorem 10.2 to the separating set K so that the resulting homeomorphism h^K of K satisfies (i) and (ii) is immediate. Now fix some $P \in \mathcal{P}_K$. If B_i, $i = 1, \ldots, m$, are the finite measured sets in the given weak neighborhood of g (see Section 11.2), choose sets $C_i^P \equiv C_i \subset B_i \cap P(K)$, $i = 1, \ldots, m$, so that $\mu\left((B_i \cap P(K)) - C_i\right)$ is very small and $\bigcup_i (C_i \cup g(C_i))$ is a relatively compact subset of $P(K)$. Hypothesis (iii) makes this possible. Let R_P be a relative n-cell in $P(K)$ which contains $\bigcup_i (C_i \cup g(C_i))$ in its interior and has boundary measure zero (using Theorem 14.6). Apply the Lusin Theorem 10.2 to any automorphism $\hat{g}$ of R_P which agrees with g on $\bigcup_i C_i$, and extend the resulting homeomorphism h^P to all of $P(K)$ by setting it equal to the identity on $P(K) - R_P$. If we piece together the homeomorphisms h^K on K, and h^P on $P(K)$ for all $P \in \mathcal{P}_K$, we obtain the required homeomorphism h. □

14.8 Induced Homeomorphism h^*

Every homeomorphism of the manifold induces a homeomorphism on the ends. We have seen in Chapter 13 two examples of these end homeomorphisms (recall the Manhattan dynamical system and the end action induced by the translation to the right homeomorphism). We now give a formal description of the homeomorphism of the ends.

Definition 14.10 *Every homeomorphism $h : X \to X$ induces a homeomorphism h^* on the ends, $h^* : E[X] \to E[X]$ defined by*

$$[h^*(e)](K) = h(e(h^{-1}(K))) \tag{14.3}$$

for all $e \in E[X]$ and compact $K \subset X$. We say that $h \in \mathcal{H}[X]$ is end preserving if h^* is the identity on $E[X]$. If h is μ-preserving (i.e., $h \in \mathcal{M}[X, \mu]$) then h^* preserves the measure μ^*. Thus every $h \in \mathcal{M}[X, \mu]$ induces a measure preserving system $(E[X], \mathcal{Q}, \mu^*, h^*)$.

Keeping in mind that the measurable sets $\mathcal{Q}$ constitute only an algebra (the algebra of clopen sets), and *not* a sigma algebra, the following definitions deserve more than the usual scrutiny.

Definition 14.11 *Let σ be a μ^*-preserving homeomorphism of $E[X]$. Then the system $(E[X], \mathcal{Q}, \mu^*, \sigma)$ is called* compressible *if there is a clopen set $Q \in \mathcal{Q}$ such that $\mu^*(\sigma Q - Q) = 0$ and $\mu^*(Q - \sigma Q) > 0$ (note since μ^* is a two-valued measure, this means that $\mu^*(Q - \sigma Q) = \infty$).*

Otherwise it is called incompressible. *The system is called* ergodic *if for every (invariant) set* $I \in \mathcal{Q}$ *with* $\mu^*(I \triangle \sigma I) = 0$, *either* $\mu^*(I) = 0$ *or* $\mu^*(E[X] - I) = 0$.

We observed earlier that if $\mu(X) < \infty$, then $\mu^*(E[X]) = 0$. Hence in this case *any* $h \in \mathcal{M}[X, \mu]$ induces an incompressible and ergodic system $(E[X], \mathcal{Q}, \mu^*, h^*)$.

In contrast to the usual case in ergodic theory where an ergodic measure preserving automorphism must necessarily be incompressible, the following example presents an ergodic and compressible end homeomorphism.

Example Let E be the totally disconnected space consisting of $-1, \ldots,$ $-1 + \frac{1}{3}, -1 + \frac{1}{2}, 0, 1 - \frac{1}{2}, 1 - \frac{1}{3}, \ldots, 1$ with its usual topology. Suppose σ fixes -1 and 1 and moves all other points to the next larger number. Assume the measure μ^* is infinite for each nonempty set. The clopen set $Q = \{0, 1 - \frac{1}{2}, 1 - \frac{1}{3}, \ldots, 1\}$ satisfies $\sigma Q - Q = \emptyset$ and $Q - \sigma Q = \{0\}$, so $(E, \mathcal{Q}, \mu^*, \sigma)$ is compressible. The only nonempty invariant clopen set is E, and hence σ is ergodic and compressible.

The reader should recognize that the translation to the right homeomorphism $\hat{h}$ on the Manhattan manifold $\hat{X}$ from the previous chapter (see Example 13.1) has an induced end homeomorphism which is topologically conjugate to the end action $(E, \mathcal{Q}, \mu^*, \sigma)$ of the previous example.

The next lemma relates the action of the induced end homeomorphism $h^* = \sigma$ on the clopen sets of ends, to the action of the homeomorphism h on the components of $X - K$ containing those ends (for sufficiently large compact sets K).

Lemma 14.12 *Let* $h \in \mathcal{M}[X, \mu]$ *and write* $\sigma = h^*$. *Let* Q_1 *and* Q_2 *belong to* $\mathcal{Q}_K$ *for some separating set* K. *Then* $\mu^*(\sigma Q_1 \cap Q_2) = \infty$ *if and only if* $\mu(h(Q_1(K)) \cap Q_2(K)) = \infty$.

Proof First suppose $\mu^*(\sigma Q_1 \cap Q_2) = \infty$. By the definition of μ^* it follows that in particular $\mu((\sigma Q_1 \cap Q_2)(B)) = \infty$ for $B = K \cup hK$.

$$
\begin{aligned}
(\sigma Q_1 \cap Q_2)(B) &= (h^* Q_1)(B) \cap Q_2(B) && \text{since } \sigma = h^* \\
&= (h Q_1 h^{-1})(B) \cap Q_2(B) && \text{by (14.3)} \\
&\subseteq h Q_1 h^{-1}(hK) \cap Q_2(K) \\
&= h(Q_1(K)) \cap Q_2(K)
\end{aligned}
$$

where the inclusion above follows from (14.1), and the fact that B contains hK and K as subsets. Therefore $\mu(h(Q_1(K)) \cap Q_2(K)) = \infty$.

Now assume $\mu^*(\sigma Q_1 \cap Q_2) < \infty$, or equivalently that for a separating set $R \supset K \cup hK$ we have $\mu((\sigma Q_1 \cap Q_2)R) < \infty$. Observe that

$$
\begin{align}
h(Q_1(K)) &= hQ_1h^{-1}(hK) \tag{14.4}\\
&= \sigma Q_1(hK). \tag{14.5}
\end{align}
$$

From (14.1) in the definition of ends, it follows that

$$
\begin{align}
(E[X] - \sigma Q_1)(hK) &= \bigcup_{e \in E[X]-\sigma Q_1} e(hK)\\
&\supset \bigcup_{e \in E[X]-\sigma Q_1} e(R) = (E[X] - \sigma Q_1)(R).
\end{align}
$$

Since the complement in X of $(E[X] - \sigma Q_1)(hK)$ is just $\sigma Q_1(hK) \cup hK$ and $(E[X] - \sigma Q_1)(R)$ has $\sigma Q_1(R) \cup R$ as its complement it follows that

$$\sigma Q_1(hK) \cup hK \subset \sigma Q_1(R) \cup R.$$

Consequently combining this with equation (14.5) from above we have

$$h(Q_1(K)) \subset \sigma Q_1(R) \cup R.$$

Since $K \subset R$, then $Q_2(K) \subset Q_2(R) \cup (R - K)$ and so

$$Q_2(K) \subset Q_2(R) \cup R.$$

Therefore

$$h(Q_1(K)) \cap Q_2(K) \subset R \cup [(\sigma Q_1)(R) \cap Q_2(R)],$$

and

$$\mu(h(Q_1(K)) \cap Q_2(K)) < \mu(R) + \mu[(\sigma Q_1 \cap Q_2)(R)] < \infty.$$

$\square$

Definition 14.13 *Suppose that $\sigma : E[X] \to E[X]$ is an end homeomorphism induced by some homeomorphism in $\mathcal{M}[X, \mu]$. Then define*

$$\mathcal{M}_\sigma[X, \mu] = \{f \in \mathcal{M}[X, \mu] : f^* = \sigma\}.$$

We can apply Baire category arguments to the space of homeomorphisms $\mathcal{M}_\sigma[X, \mu]$ because of the following lemma.

Lemma 14.14 $\mathcal{M}_\sigma[X, \mu]$ *is a closed subset of $\mathcal{M}[X, \mu]$ with respect to the compact-open topology.*

Proof Denote by Ψ, the * map; i.e.,

$$\Psi : \mathcal{M}[X,\mu] \to \mathcal{M}[E[X],\mu^*]$$

defined by $\Psi(h) = h^*$. Since the * map Ψ is continuous, the inverse image of a point $\sigma \in \mathcal{M}[E[X],\mu^*]$ is closed in $\mathcal{M}[X,\mu]$. But this inverse image is nothing more than $\mathcal{M}_\sigma[X,\mu]$. $\qquad\square$

Recall that when we considered the Manhattan dynamical system $(\hat{X}, \hat{\mu}, \hat{h})$, whose induced end homeomorphism $\hat{h}^*$ is compressible, we showed that $\hat{h}$ could not be $\hat{\mu}$-recurrent. We now generalize this to show for all (X,μ) that *any* homeomorphism $f \in \mathcal{M}[X,\mu]$ which induces a compressible end homeomorphism $f^* : E[X] \to E[X]$ cannot be μ-recurrent on (X,μ).

Lemma 14.15 *If $(E[X], \mathcal{Q}, \mu^*, \sigma)$ is a compressible system then no homeomorphism in $\mathcal{M}_\sigma[X,\mu]$ is μ-recurrent, and therefore no homeomorphism in $\mathcal{M}_\sigma[X,\mu]$ can be ergodic.*

Proof Suppose that h belongs to $\mathcal{M}_\sigma[X,\mu]$ (i.e., the action of h on the ends is $h^* = \sigma$). Then because σ is compressible, there is some clopen set of ends $Q \in \mathcal{Q}$ which satisfies $\mu^*(\sigma Q - Q) = 0$ and $\mu^*(Q \cap \sigma(E[X] - Q)) > 0$. Choose a separating set K such that $Q \in \mathcal{Q}_K$. Then by Lemma 14.12 $\mu(h(Q(K)) - Q(K)) < \infty$ and $\mu(Q(K) - h(Q(K))) = \infty$. But taking $V = Q(K)$, this yields

$$\mu(h(V) - V) \neq \mu(V - h(V))$$

violating the condition in Lemma 13.3 necessary for h to be μ-recurrent. $\qquad\square$

In order to state our next two theorems, we will need to define some properties of finite $m \times m$ matrices $\mathbf{T} = (t_{i,j})$ whose entries are all 0s and 1s. Such a matrix $\mathbf{T}$ is called *irreducible* if for any i and j there is a positive integer a such that $t_{i,j}^a = 1$. If moreover there is a single positive integer N such that $t_{i,j}^N = 1$ for all i and j, then $\mathbf{T}$ has the stronger property that we will call *mixing*. The matrix $\mathbf{T}$ is called *recurrent* if for every i there is a positive integer a such that $t_{i,i}^a = 1$. The greatest common divisor of such powers a is called the *period* of i. (If $\mathbf{T}$ is irreducible then every i has the same period.) If every i has period 1 then we say that $\mathbf{T}$ is *aperiodic*. It is easy to see that $\mathbf{T}$ is mixing if and only if it is both irreducible and aperiodic, and that if $\mathbf{T}$ is recurrent it can be decomposed into irreducible submatrices. Similar definitions are given in Appendix 1 for stochastic matrices with either finitely or infinitely many states.

The next theorem gives a kind of 'ergodic decomposition' for an incompressible end homeomorphism σ with respect to a given clopen partition of E^∞. It produces a minimal 'clumped' partition of E^∞ into σ-invariant clopen sets, of which the given partition is a refinement. This result will be used in the proof of Theorem 15.1 (in Lemma 16.3) to produce an ergodic μ-preserving manifold homeomorphism h with $h^* = \sigma$.

Theorem 14.16 *Let $\sigma : E \to E$ be an incompressible end homeomorphism and let $E_1, \ldots, E_m$ form a partition of E into clopen sets of infinite μ^* measure. Define an $m \times m$ 0-1 matrix $\boldsymbol{T}$ by $t_{ij} = 1$ if $\mu^*(\sigma E_i \cap E_j) = \infty$ and $t_{ij} = 0$ otherwise. Then the relation $E_i \sim E_j$ if $t_{ij}^a > 0$ for some integer $a \geq 1$ is an equivalence relation on the set $\{E_1, \ldots, E_m\}$, and hence $\boldsymbol{T}$ is recurrent. The corresponding equivalence classes determine a partition of E^∞ into minimal σ-invariant clopen sets $C_1, \ldots, C_p$ (which are each unions of the sets E_i) and an associated decomposition of $\boldsymbol{T}$ into square irreducible submatrices.*

Proof Consider the directed graph G with vertex set $\{1, \ldots, m\}$ and an arc from i to j if $t_{ij} = 1$. Since σ preserves the measure μ^* it follows that every vertex has at least one arc going out of it. First suppose that the relation $\sim$ is not symmetric. Then for some vertices i and j there is a path from i to j but no path from j to i. Relabel the vertices so that $i = 1$, $j = b$, and $1, 2, \ldots, b$ is such a path (from i to j). If there is no path from b to any of the vertices $1, 2, \ldots, b$, set $c = b$; otherwise let c be the least number (under the relabeling) such that there is a path from b to c. Note that in either case $c > 1$ by assumption (no path from j to i). Let F denote the set of vertices z such that there is a path from c to z. Observe that $c - 1$ does not belong to F since in this case the path from b to c followed by the path from c to $c - 1$ would be a path from b to $c - 1$, contradicting the minimality of c. Now define the clopen end set $Q = E_{c-1} \cup \left(\bigcup_{k \in F} E_k \right)$ and observe that $\sigma(Q) = \bigcup_{k \in F} E_k \subset Q$ and $\mu^*(Q - \sigma Q) = \mu^*(E_{c-1}) = \infty$. This would imply that σ is compressible, contrary to hypothesis, so our additional assumption that the relation $\sim$ is not symmetric was false. The transitivity of the relation $\sim$ is obvious. Reflexivity follows from the fact that for each vertex k there is an arc to some vertex k' (possibly equal to k) and consequently by symmetry there is a path from k' back to k. $\qquad\square$

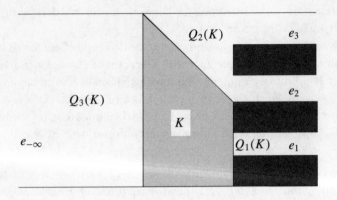

Fig. 14.2. The four-ended cylinder

We illustrate these concepts with the following example.

Example 14.17 (Four-ended cylinder) *Define a manifold by*

$$X = R \times [0,1] - \bigcup_{k=0}^{2} (1,\infty) \times \left(\frac{2k}{6}, \frac{2k+1}{6} \right),$$

with the top and bottom identified, that is, $(x,0)$ with $(x,1)$ for all $x \in R$. This manifold has four ends, all of infinite measure, which we denote by $E = \{e_{-\infty}, e_1, e_2, e_3\}$. Roughly speaking $e_{-\infty}$ corresponds to going to infinity along the line $(x,1/2)$ as x goes to $-\infty$, while each end e_i, $i = 1,2,3$ corresponds to going to infinity along the line $\left(x, \frac{4i-1}{12} \right)$ as x goes to ∞. The homeomorphism defined by $h(x,y) = (x, y+1/3)$, where addition in $y + 1/3$ is taken mod 1, preserves area. The induced end homeomorphism h^* fixes $e_{-\infty}$ and cyclically permutes the remaining ends $e_1 \to e_2 \to e_3 \to e_1$. It is incompressible. Let K denote the compact set bounded by the circle $\{-1\} \times [0,1]$ and the lines joining the points $(-1,0), (+1,0), (+1,1/2)$ and $(-1,1) = (-1,0)$. The set $X - K$ has three components, corresponding to the end partition $\mathcal{P}_K = \{Q_1 = \{e_1\}, Q_2 = \{e_2, e_3\}, Q_3 = \{e_{-\infty}\}\}$. The incidence matrix $\mathbf{T} = \mathbf{T}(h, K)$ defined by $t_{ij} = 1$ if $h^*(Q_i) \cap Q_j \neq \emptyset$ and otherwise 0, and the submatrix $\mathbf{B}$ corresponding to the h^* invariant set $C = \{Q_1, Q_2\}$ are *given by*

$$\mathbf{T} = \begin{pmatrix} 0 & 1 & 0 \\ 1 & 1 & 0 \\ 0 & 0 & 1 \end{pmatrix} \quad and \quad \mathbf{B} = \begin{pmatrix} 0 & 1 \\ 1 & 1 \end{pmatrix}.$$

If we want the matrix $\mathbf{T}$ in Theorem 14.16 to be mixing rather than merely irreducible, we must make stronger assumptions on σ. The required notion is a variation on Furstenberg's definition of topological weak mixing [64], which we call componentwise weak mixing.

Definition 14.18 *A homeomorphism σ of a compact topological space into itself is called* componentwise weak mixing *if for all clopen sets U, V the set $\{k : \sigma^k U \cap V \neq \emptyset\}$ contains consecutive integers.*

Theorem 14.19 *Let $h \in \mathcal{M}_\sigma[X, \mu]$ and assume $(E[X], \mathcal{Q}, \mu^*, \sigma)$ is incompressible and σ is componentwise weak mixing on the invariant subset $E^\infty[X]$ consisting of the ends of infinite measure. For any separating set K, let $E_1, \ldots, E_m$ be an enumeration of the elements of $\mathcal{P}_K$ having infinite μ^*-measure. Then the 0–1 matrix $\mathbf{T}$, defined by $t_{ij} = 1$ if and only if $\mu\left(h\left(E_i(K)\right) \cap E_j(K)\right) = \infty$, is mixing.*

Proof For $i = 1, \ldots, m$, define $E_i^\infty = E_i \cap E^\infty[X]$, and observe that these are clopen sets in the relative topology on $E^\infty[X]$. It follows from Lemma 14.12 that $\sigma^k(E_i^\infty) \cap E_j^\infty \neq \emptyset$ implies that $t_{ij}^k = 1$. The assumption of componentwise weak mixing implies that the matrix $\mathbf{T}$ is irreducible, so that all states $i = 1, \ldots, m$ have the same period p. (This means that $t_{ii}^k = 1$ only when p divides k.) Consequently $\{k : \sigma^k(E_i^\infty) \cap E_i^\infty \neq \emptyset\} \subset \{k : t_{ii}^k = 1\} \subset p\mathbb{Z}$, the multiples of p. If $\mathbf{T}$ is not aperiodic, then $p > 1$ and the set $p\mathbb{Z}$ does not contain two consecutive integers. Thus as $\mathbf{T}$ is not aperiodic, then σ could not be componentwise weak mixing, contrary to assumption. It follows that $\mathbf{T}$ is irreducible and aperiodic and consequently mixing. $\qquad\square$

14.9 The Charge Induced by a Homeomorphism

Every homeomorphism $h \in \mathcal{M}[X, \mu]$ induces on the ends $E = E[X]$ a charge (a finitely additive signed measure) $c = c_h$ defined on the subalgebra $\mathcal{I}_{h^*}$ of $\mathcal{Q}$ consisting of h^*-invariant clopen sets. This section is devoted to defining this end charge c and establishing some of its elementary properties.

In the two examples in Chapter 13, the unit translation on the strip dynamical system $(\check{X}, \check{\mu}, \check{h})$, and the Manhattan dynamical system $(\hat{X}, \hat{\mu}, \hat{h})$, we computed the quantity

$$\mu(V - h(V)) - \mu(h(V) - V)$$

where V was an unbounded component containing the end fixed at $+\infty$ by both of the systems. The fact the the above quantity was nonzero was used to prove that both of the dynamical systems $(\hat{X}, \hat{\mu}, \hat{h})$ and $(\check{X}, \check{\mu}, \check{h})$ are not μ-recurrent (see Lemma 13.3). These examples and the quantity $\mu(V - h(V)) - \mu(h(V) - V)$ (which we encountered in the previous section as well) motivate the definition of the charge of a homeomorphism c_h for a general $h \in \mathcal{M}[X, \mu]$.

Fix a homeomorphism $h \in \mathcal{M}[X, \mu]$ and let σ denote the action of h^* on the ends. Denote by $\mathcal{I} = \mathcal{I}_\sigma$ the subalgebra of σ-invariant clopen sets in $\mathcal{Q}$ given by $\mathcal{I} = \{I \in \mathcal{Q} : \mu^*(I \bigtriangleup \sigma I) = 0\}$. For every $I \in \mathcal{I}$ there is a separating set $K \subset X$ such that $I \in \mathcal{Q}_K$. For such a set, define

$$c(I, K) = \mu(I(K) - h(I(K))) - \mu(h(I(K)) - I(K)). \qquad (14.6)$$

This quantity measures the net flow of mass by h into the invariant set of ends in $I(K)$.

In the definition of $c(I, K)$ above, we first note that this difference is well defined. To see this we note that $I \in \mathcal{I}$ is an invariant clopen set, and so $\mu^*(I - \sigma I) = 0 = \mu^*(\sigma I - I)$; thus from Lemma 14.12, we have $\mu(I(K) - h(I(K))) < \infty$ and $\mu(h(I(K)) - I(K)) < \infty$. Consequently the difference in (14.6) is well defined. The fact that the end charge does not depend on the set K follows from

Lemma 14.20 *Suppose I is a clopen set of ends invariant under $\sigma = h^*$ belonging to both $\mathcal{Q}_K$ and $\mathcal{Q}_{K'}$. Then $c(I, K) = c(I, K')$, a number which will be simply written as $c(I)$.*

Proof Let R be any separating set containing $K \cup hK \cup h^{-1}K$. Such a separating set can be found using Theorem 14.6. We will show $c(I, R) = c(I, K)$. This will prove the lemma since by Theorem 14.6 we can always find an R which simultaneously has this relationship to both K and K'. Let $B = I(K) - I(R)$. Since $R \supset K$, then it follows from (14.1) that

$I(R) \subset I(K)$ and $I(K) = (R \cap I(K)) \cup I(R)$. Consequently

$$B = I(K) - I(R) = R \cap I(K)$$

is a subset of a compact set and therefore has finite measure. If we apply equation (13.2) from Lemma 13.4 to the finite measured set $V = B$, and the μ-preserving automorphism $g = h$, then we get

$$\mu(h(B) - B) - \mu(B - h(B)) = 0.$$

Now observe that

$$h(B) - B = [I(R) - h(I(R))] \bigcup_{\text{disj}} [h(I(K)) - I(K)]$$

$$B - h(B) = [I(K) - h(I(K))] \bigcup_{\text{disj}} [h(I(R)) - I(R)].$$

Therefore $c(I, R) - c(I, K) = \mu(h(B) - B) - \mu(B - h(B)) = 0.$ $\square$

As a consequence of this lemma we can define the *charge* $c = c_h$ *induced by a homeomorphism* $h \in \mathcal{M}[X, \mu]$ on the algebra $\mathcal{I}$ of h^*-invariant clopen sets of ends in $E = E[X]$, by equation (14.6). We note that if $\sigma = h^*$ is ergodic on (E, μ^*), then there are no nontrivial invariant sets of ends and $\mathcal{I}_\sigma = \{E, \emptyset\}$; in this case the charge is identically zero $(c_h(E) = 0$, and $c_h(\emptyset) = 0)$. More generally the following lemma gives further properties of the charge.

Lemma 14.21 *Let* $c = c_h$ *be the end charge induced by some homeomorphism* h *in* $\mathcal{M}[X, \mu]$ *on the algebra* $\mathcal{I}$ *of* h^*-*invariant clopen sets of* E. *Then*

(i) $c(E) = 0$.

(ii) *If* $\mu^*(I) = 0$ *then* $c(I) = 0$.

(iii) *If* $I_1, I_2 \in \mathcal{I}$ *with* $I_1 \cap I_2 = \emptyset$, *then* $c(I_1 \cup I_2) = c(I_1) + c(I_2)$.

(iv) *Let* σ *be a fixed* μ^*-*preserving homeomorphism of the ends. The charge* $c = c_h$ *is continuous on each space* $\mathcal{M}_\sigma[X, \mu]$ *in the sense that for each* $I \in \mathcal{I}_\sigma$, *the function* $h \to c_h(I)$ *is continuous on* $\mathcal{M}_\sigma[X, \mu]$.

Proof (i) For any separating set K, we have $E \in \mathcal{Q}_K$ and $E(K) = X - K$. By Lemma 13.4 and equation (13.2) applied to the finite measured set $V = K$, and h

$$c(E) = c(E, K) = \mu(h(K) - K) - \mu(K - h(K)) = 0.$$

(ii) If $\mu^*(I) = 0$ then by definition there is a separating set K with $\mu(I(K)) < \infty$. Again applying Lemma 13.4, this time to the finite measured set $V = I(K)$, it follows that

$$c(I) = \mu(I(K) - h(I(K))) - \mu(h(I(K)) - I(K)) = 0.$$

(iii) For any K with $I_1, I_2 \in \mathcal{Q}_K$, because $I_1 \cap I_2 = \emptyset$, we have $I_1(K) \cap I_2(K) = \emptyset$. For such a K, $c(I_1 \cup I_2, K)$ equals $c(I_1, K) + c(I_2, K)$.

(iv) Fix $I \in \mathcal{I}_\sigma$ and let

$$\Phi : \mathcal{M}_\sigma[X, \mu] \to R^1$$

be the map $\Phi(h) = c_h(I)$. We prove the stronger assertion that $\Phi(\cdot)$ is in fact continuous in the weak topology on $\mathcal{M}_\sigma[X, \mu]$. Using Lemma 14.20, fix any compact set K such that I belongs to $\mathcal{I} \cap \mathcal{Q}_K$, so that $c(I) = c(I, K)$. If $h_j \to h$ in the weak topology on $\mathcal{M}_\sigma[X, \mu]$, then

$$\mu(I(K) - h_j(I(K))) \quad \to \quad \mu(I(K) - h(I(K))) \tag{14.7}$$
$$\mu(h_j(I(K)) - I(K)) \quad \to \quad \mu(h(I(K)) - I(K)) \tag{14.8}$$

so that $\Phi(h_j) = c_{h_j}(I, K) \to \Phi(h) = c_h(I, K)$ as required. $\qquad\square$

Note The proof above actually shows that equations (14.7) and (14.8) are valid if $h_j \to h$ in the weak topology in $\mathcal{M}[X, \mu]$, which is a stronger result than that required by the lemma.

We observe that in a trivial sense the finite additivity of c on $\mathcal{I}$ can be extended to countable additivity since if a clopen set $I \in \mathcal{I}$ is the denumerable disjoint union of clopen sets $I_1, I_2, \ldots$, then the compactness of I implies all but a finite number of those sets must be the empty set. We now give simple computations of the charge induced by some homeomorphisms.

Examples

(i) *The charge $c_{\hat{h}}$ for the Manhattan dynamical system $(\hat{X}, \hat{\mu}, \hat{h})$*: Since the action $\hat{h}^*$ on $E[\hat{X}] = \{-\infty, \ldots, -1, 0, 1, \ldots, +\infty\}$ is ergodic, there are no nontrivial invariant clopen sets and so $c_{\hat{h}} \equiv 0$. We note however that the action of $\hat{h}^*$ is *compressible*.

(ii) *The charge $c_{\check{h}}$ for the unit translation on the strip dynamical system $(\check{X}, \check{\mu}, \check{h})$*: Here the end set is $E[\check{X}] = \{-\infty, +\infty\}$ and $\check{h}$ fixes both of these ends. Thus if we let $I_- = \{-\infty\}$ and $I_+ = \{+\infty\}$ be the two nontrivial $\check{h}^*$-invariant clopen sets then for $K = [3/8, 5/8] \times [0, 1]$ we computed in (13.4) with $V = I_+(K)$ that

$$c_{\check{h}}(I_+) = c_{\check{h}}(I_+, K) = 1.$$

Consequently $c_{\bar{h}}(I_-) = -1$.

(iii) *If (X, μ) has no ends of infinite measure then for any $h \in \mathcal{M}[X, \mu]$ the charge $c_h(I) = 0$ for all invariant clopen sets $I \in \mathcal{I}_{h^*}$. This follows easily from part (ii) of the above lemma because the finiteness of μ implies that $\mu^*(I) = 0$ for every I.*

(iv) *If (X, μ) has exactly one end of infinite measure then for any $h \in \mathcal{M}[X, \mu]$ the charge $c_h(I) = 0$ for all invariant clopen sets $I \in \mathcal{I}_{h^*}$. If I contains only ends of finite measure, then $\mu^*(I) = 0$ and so again (ii) of Lemma 14.21 implies $c_h(I) = 0$. If I contains the end of infinite measure, then $E[X] - I$ has only ends of finite measure and so $c_h(E[X] - I) = 0$. The finite additivity of c_h along with $c_h(E[X]) = 0$ implies that $c_h(I) = 0$.*

A consequence of the next result shows that the end charge varies continuously on $\mathcal{M}_\sigma[X, \mu]$, a strengthened form of Lemma 14.21(iv).

Theorem 14.22 *For a fixed homeomorphism σ of the ends, let $\mathcal{M}_\sigma[X, \mu]$ be the homeomorphisms which induce the end homeomorphism σ, and let $\mathcal{I} = \mathcal{I}_\sigma$ denote the algebra of σ-invariant clopen subsets of $E[X]$.*

For any fixed end charge c on $\mathcal{I}_\sigma$, the set $\mathcal{M}_\sigma^c[X, \mu]$, consisting of all homeomorphisms $h \in \mathcal{M}[X, \mu]$ with $h^* = \sigma$ and fixed end charge $c_h = c$, is closed in $\mathcal{M}_\sigma[X, \mu]$. In particular, the set $\mathcal{M}_\sigma^0[X, \mu]$ consisting of all homeomorphisms $h \in \mathcal{M}[X, \mu]$ with $h^* = \sigma$ and identically zero end charge is a closed subset of the topologically complete space $\mathcal{M}[X, \mu]$ with the compact-open topology.

Furthermore, $\mathcal{M}_\sigma^c[X, \mu]$ is invariant under right composition with any $h \in \mathcal{M}[X, \mu]$ of compact support.

Proof First fix an end charge c on $\mathcal{I} = \mathcal{I}_\sigma$. Since

$$\mathcal{M}_\sigma^c[X, \mu] = \bigcap_{I \in \mathcal{I}} \Phi^{-1}\{c(I)\}$$

and the continuity of Φ (from Lemma 14.21) implies each set in the intersection is closed in $\mathcal{M}_\sigma[X, \mu]$ with the compact-open topology, it follows that $\mathcal{M}_\sigma^c[X, \mu]$ is also closed in $\mathcal{M}[X, \mu]$.

Next suppose that $h_1 \in \mathcal{M}_\sigma[X, \mu]$ has end charge $c_{h_1} = c$. If $h_2 \in \mathcal{M}[X, \mu]$ has compact support, then $h_2(R) = R$ and $h_2(I(R)) = I(R)$ for some compact R containing the support of h_2, with $I \in \mathcal{Q}_R$. Then

$c_{h_1 h_2}(I) = c_{h_1 h_2}(I, R)$, and

$$
\begin{aligned}
c_{h_1 h_2}(I, R) &= \mu(I(R) - h_1 h_2(I(R))) - \mu(h_1 h_2(I(R)) - I(R)) \\
&= \mu(I(R) - h_1(I(R))) - \mu(h_1(I(R)) - I(R)) \\
&= c_{h_1}(I, R) \\
&= c_{h_1}(I).
\end{aligned}
$$

$\square$

The first part of this theorem will enable us to carry out Baire category proofs in $\mathcal{M}_\sigma^0[X, \mu]$, the zero charge homeomorphisms with end action σ. The case of nonzero charge is dealt with in the following.

Theorem 14.23 *If $h \in \mathcal{M}[X, \mu]$ induces a nonzero charge c_h, then h is not the compact-open limit of ergodic or even μ-recurrent homeomorphisms in $\mathcal{M}[X, \mu]$.*

Proof Since c_h is not identically zero we have $c_h(I) \neq 0$ for some h^*-invariant clopen set of ends I. This set I belongs to $\mathcal{Q}_K$ for some compact separating set $K \subset X$. Setting $V = I(K)$ we have

$$
c_f(I) = c_f(I, K) = \mu(f(V) - V) - \mu(V - f(V)) \neq 0
$$

not only for $f = h$ but also for all f in some compact-open neighborhood of h, by the continuity of $c_f(I)$ in f (Theorem 14.22). It follows that for all f in this neighborhood, f is not recurrent (and hence not ergodic) by Lemma 13.3. $\square$

14.10 h-moving Separating Sets

In the following chapters we will show how to approximate a given homeomorphism h by an ergodic automorphism f which is close on a compact separating set K. The construction we will give has the property that not only is f uniformly close to h on K but it also satisfies

$$
f(Q(K)) = h(Q(K)) \text{ for every end set } Q \in \mathcal{Q}_K.
$$

The exact statement of this approximation result is given in Chapter 16 as Lemmas 16.2 and 16.3. Note that the displayed property above implies that in addition we have $f(K) = h(K)$. Consider first the very special case where h is the identity homeomorphism, and observe that any f satisfying these conditions would have K and all the sets $Q(K)$,

$Q \in \mathcal{Q}_K$, as nontrivial invariant sets, and hence could not be ergodic. This observation holds more generally if the algebra $\mathcal{A}$ generated by the partition of X

$$X = K \cup E_1(K) \cup E_2(K) \cup \cdots \cup E_r(K),$$

where $\mathcal{P}_K = \{E_i(K)\}_{i=1}^r$, has an atom A which is periodic in $\mathcal{A}$ under h (this means that for some period $m \geq 1$, we have $h^j(A) \in \mathcal{A}$, $j = 0, \ldots, m$, with $h^m(A) = A$), since the set $\bigcup_{j=0}^m h^j(A)$ would be invariant. So in order to carry out the ergodic approximation we will need some assumption regarding the behavior of h on the sets $Q(K)$ which precludes the existence of an h-periodic atom in $\mathcal{A}$. This motivates the following definition.

Definition 14.24 *Let $h \in \mathcal{M}[X, \mu]$ be given. A separating set K is called h-moving if $\mu(h(K) \cap K) > 0$ and $\mu(h(I(K)) \cap K) > 0$ for every h^*-invariant clopen set of ends $I \in \mathcal{Q}_K$.*

This concept is applied in the following lemma.

Lemma 14.25 *Let K be an h-moving separating set. Then the algebra $\mathcal{A}$ generated by the partition of X*

$$X = K \cup E_1(K) \cup E_2(K) \cup \cdots \cup E_r(K),$$

where $\mathcal{P}_K = \{E_i(K)\}_{i=1}^r$, does not have an atom A which is periodic in $\mathcal{A}$ under h.

Proof Suppose that an atom $A \in \mathcal{A}$ has h-period m in $\mathcal{A}$. If $A = K$ then since $h(K)$ cannot be any of the sets $E_i(K)$, we must have $m = 1$ and $h(K) = K$, which violates the h-moving assumption.

If $A = E_i(K)$ for some i, then the h-invariant set $\bigcup_{j=0}^m h^j(A)$ equals $I(K)$ for the h^*-invariant set $I \in \mathcal{Q}_K$ containing E_i. This means that $I(K) = h(I(K))$ belongs to the algebra $\mathcal{A}$. However, it follows from the definition of h-moving that

$$0 < \mu(h(I(K)) \cap K) < \mu(K)$$

and consequently $h(I(K))$ cannot belong to the algebra $\mathcal{A}$. $\qquad\square$

Of course the compact-open topology has been defined on $\mathcal{G}[X, \mu]$ so that the relative topology on $\mathcal{M}[X, \mu]$ is the usual compact-open topology (uniform convergence on compact sets). For this relative topology we can require that the compact set K, which defines the basic compact-open neighborhood $\mathcal{C}(h, K, \epsilon)$ of a homeomorphism $h \in \mathcal{M}[X, \mu]$ (see Section 11.2), be a separating set; i.e., K is a relative n-cell with $\mu(\text{Bdry } K) = 0$ and $X - K$ has no unbounded components. Furthermore, the following lemma shows that we can require K in the compact-open neighborhood above to be an h-moving set.

Lemma 14.26 *Let* $h \in \mathcal{M}[X, \mu]$, $\epsilon > 0$, *and a separating set* K *be given. Then there is an* $h' \in \mathcal{M}[X, \mu]$ *with compact support such that* $\sup_{x \in X} d(x, h'(x)) < \epsilon$ *and* K *is* hh'-moving. *Consequently there is a subbasic family of compact-open topology open sets of the form* $\mathcal{C}(g, K, \epsilon)$, $g \in \mathcal{M}[X, \mu]$, *where* K *is* g-moving.

Proof For each h^*-invariant set of ends I in $\mathcal{I}_{h^*} \cap \mathcal{Q}_K$ choose distinct topological n-balls B_I with diameter less than ϵ such that $\mu(B_I \cap I(K)) > 0$ and $\mu(B_I \cap K) > 0$. For each I choose a μ-preserving homeomorphism h_I in $\mathcal{M}[X, \mu]$ with support in B_I such that $B_I \cap I(K)$ is not invariant under h_I. There are several ways of constructing h_I. One way is to take h_I to be any μ-preserving ergodic homeomorphism of B_I which fixes the boundary of B_I – extend h_I to X by setting it to be the identity off B_I. Such an ergodic homeomorphism exists by the results of Part I. The homeomorphism h' is the composition of these finitely many h_I's. By our construction h' is small and K is $g = hh'$-moving. $\square$

14.11 End Conditions for Homeomorphic Measures

For noncompact manifolds X, we may ask when two OU measures μ and ν on X are homeomorphic (i.e., $\mu = \nu h$ for some homeomorphism h of X). Complications can arise if the behavior of the measures at the ends of the manifold is not taken into account. For example area measure μ on the strip $\check{X}$, which is infinite on both ends, cannot be homeomorphic to any measure ν which is finite on one end of the strip but infinite on the other. R. Berlanga and D. Epstein showed [38] in 1981 that a sufficient condition for two sigma finite OU measures on a connected sigma compact manifold X to be homeomorphic is if the measures are infinite on the same set of ends.

Theorem A2.8 [Berlanga–Epstein] *Let* X *be a sigma compact, connected* n-manifold $(n \geq 2)$ *and let* μ *and* ν *be two nonatomic Borel measures, positive on open sets, finite on compact sets and zero on the* *boundary of X (i.e., two OU measures).   Then μ and ν are homeomorphic if $\mu(X) = \nu(X)$ and μ and ν are infinite on the same set of ends (i.e., the measures μ^* and ν^* induced on the ends are identical). The homeomorphism h of X such that $\nu = \mu h$ can be chosen to fix the boundary of X (i.e., $h \in \mathcal{H}[X, \partial X]$), and be end preserving.*

Further discussion and a proof of this theorem is given in the second Appendix.

15

Ergodic Homeomorphisms: The Results

15.1 Introduction

In this chapter we determine necessary and sufficient conditions for a measure preserving homeomorphism h of a sigma compact manifold X to be the limit of ergodic homeomorphisms, in the compact-open topology. We have already shown in Prasad's Theorem 12.4 that such an approximation is *always* possible (for any h) when X is Euclidean space R^n with Lebesgue measure. In Examples 13.1 and 13.2 we showed on the contrary that when h is the unit translation on either the Manhattan manifold or the strip manifold, such an approximation is *not* possible. The obstruction to an ergodic approximation for these systems was explained in terms of the ends E of the manifold X in Chapter 14: The unit translation on the Manhattan manifold induces a homeomorphism on the ends which is compressible, and hence ergodic approximation is precluded by Lemma 14.15; the unit translation on the strip manifold is incompressible, but since it induces a nonzero charge on the ends, an ergodic approximation is ruled out by Theorem 14.23. Conditions on the ends will be used in this chapter to obtain *positive* results on ergodic approximation. The results in this chapter come from [20] and [22].

We will obtain complete answers to a number of simple questions regarding ergodic approximation in $\mathcal{M}[X, \mu]$ with respect to the compact-open topology. Since the G_δ-ness of ergodicity in $\mathcal{M}[X, \mu]$ follows immediately from the similar result (Theorem 11.1) for automorphisms with respect to the weak topology, denseness and genericity of ergodicity are equivalent, so we phrase our questions in terms of genericity. The first question asks for which manifolds (X, μ) is ergodicity generic in *the full space* $\mathcal{M}[X, \mu]$. That is, how far can we generalize Theorem 12.4 from R^n to a general manifold (X, μ)? The answer, given in Corollary 15.3, is that ergodicity is generic in $\mathcal{M}[X, \mu]$ if and only if (X, μ) has at most one end of infinite measure. This covers all compact manifolds (and hence implies the Oxtoby–Ulam Theorem, Theorem 7.1), all manifolds with finite measure, and Euclidean space with Lebesgue measure (Theorem 12.4). In the case that (X, μ) has two or more ends of infinite measure, it follows that the most we can hope for is that ergodicity is generic in *some component* $\mathcal{M}_\sigma[X, \mu]$ determined by an end homeomorphism σ. What is the condition we must put on σ? The answer, given in Corollary 15.4, is that σ must be incompressible and ergodic with respect to the induced end measure μ^*. For those who prefer a purely topological condition sufficient for generic ergodicity in $\mathcal{M}_\sigma[X, \mu]$, Corollary 15.7 shows that the topological transitivity of σ on the ends E is such a condition. Finally, we may ask, in the case that ergodicity is not generic in $\mathcal{M}_\sigma[X, \mu]$, does there *exist* an ergodic homeomorphism in $\mathcal{M}_\sigma[X, \mu]$, that is, with induced end homeomorphism σ? The answer to this question is yes, if and only if σ is incompressible (Corollary 15.5). The above results may also be summarized by saying that h is the compact-open topology limit of ergodic homeomorphisms if and only if its induced end homeomorphism h^* is incompressible and its induced end charge c_h is identically zero (Corollary 15.9). This is the result described (in Section 13.3) as Theorem F.

The reader may be wondering why all the results from this chapter mentioned above are called 'corollaries'. This is because we have organized the proofs so that all these diverse observations are consequences of a single result, which we state below as Theorem 15.1. (A similar but easier finite measure analog is also stated as Theorem 15.2.) This chapter shows how all the answers to the questions of ergodic approximation and ergodic genericity follow from Theorem 15.1. The proof of Theorem 15.1 will be carried out in Chapter 16, together with a similar proof for Theorem 15.2.

Theorem 15.1 *Let μ be an OU measure on a sigma compact manifold X, and let μ^* be the induced 0–∞ measure on the ends E of X. Let $\sigma : E \to E$ be an induced end homeomorphism such that the system $(E, \mathcal{Q}, \mu^*, \sigma)$ is incompressible, and let $\mathcal{M}_\sigma^0[X, \mu]$ denote the set of μ-preserving homeomorphisms h of X which induce the end homeomorphism σ (that is, $h^* = \sigma$) and have identically zero charge. Then the ergodic homeomorphisms in $\mathcal{M}_\sigma^0[X, \mu]$ form a dense G_δ subset of $\mathcal{M}_\sigma^0[X, \mu]$, with respect to the compact-open topology.*

In proving this theorem in the next chapter we will find that in the special case when $\mu(X) < \infty$ a much stronger result is true: namely, ergodicity $\mathcal{E}$ can be replaced by any other measure theoretic property $\mathcal{F}$ which is typical in $\mathcal{G}[X, \mu]$ with respect to the weak topology. Note that when the measure of X is finite, there are no ends of infinite measure and consequently every μ-preserving homeomorphism of X induces an incompressible end homeomorphism. Furthermore, when $\mu(X) < \infty$ all ends have finite measure and the end homeomorphism h^* is necessarily ergodic on $(E, \mathcal{Q}, \mu^*)$ for all homeomorphisms $h \in \mathcal{M}[X, \mu]$. Thus even in the case of noncompact manifolds, when $\mu(X) < \infty$ the situation is very much like the case for compact manifolds (at least as regards typical measure theoretic properties).

Theorem 15.2 *Let μ be a finite OU measure on a sigma compact manifold X. Then for any conjugate invariant property which defines a dense G_δ subset $\mathcal{F}$ of $\mathcal{G}[X, \mu]$ with respect to the weak topology, $\mathcal{F} \cap \mathcal{M}[X, \mu]$ is a dense G_δ subset of $\mathcal{M}[X, \mu]$.*

15.2 Consequences of Theorem 15.1

We now state and prove various consequences of Theorem 15.1. In all of these results μ denotes an OU measure on a sigma compact manifold X. The first consequence of Theorem 15.1 answers the question: For which measured manifolds (X, μ) is ergodicity typical in the space of *all* μ-preserving homeomorphisms?

Corollary 15.3 *The ergodic μ-preserving homeomorphisms of (X, μ) form a compact-open dense G_δ subset of $\mathcal{M}[X, \mu]$ if and only if the measured manifold (X, μ) has at most one end of infinite measure.*

Furthermore if the manifold has two or more ends of infinite measure then there is an open set of homeomorphisms, none of which is μ-recurrent.

Note that this corollary subsumes our two earlier results stating that ergodicity is typical in $\mathcal{M}[X, \mu]$ when X is the unit cube I^n (Theorem 7.1) or Euclidean space R^n (Theorem 12.5). In the former case there are *no* ends (this applies to any compact manifold) while in the latter case there is exactly *one* end. The measure property of the ends becomes relevant when there are at least two ends, as in the strip manifold $\check{X} = R^1 \times [0, 1]$. If the given measure μ on $\check{X}$ gives both ends infinite measure (area measure λ is one such measure), then the above Corollary says that ergodicity is *not* typical. A proof of this special case has already been given in Theorem 13.6. However, for any measure on the strip $\check{X}$ which gives infinite measure to only one end, the corollary says that ergodicity *is* typical. One such measure is obtained by the product $e^t dt \times dt$, which gives infinite measure to the right end only.

Proof of Corollary 15.3 Suppose that (X, μ) has exactly one end of infinite measure. If σ is any homeomorphism of the ends such that $\mathcal{M}_\sigma[X, \mu] \neq \emptyset$, then σ fixes the single end of infinite measure. It follows from the computation of charge, in Example (iv) after Lemma 14.21 in the previous chapter, that σ is incompressible and has zero charge, so that $\mathcal{M}_\sigma[X, \mu] = \mathcal{M}_\sigma^0[X, \mu]$. Thus by Theorem 15.1, $\mathcal{E} \cap \mathcal{M}_\sigma[X, \mu]$ is compact-open dense G_δ in every component $\mathcal{M}_\sigma[X, \mu]$ of $\mathcal{M}[X, \mu]$. Since $\mathcal{E} \cap \mathcal{M}[X, \mu] = \bigcup_\sigma (\mathcal{E} \cap \mathcal{M}_\sigma[X, \mu])$, where the union is the disjoint union taken over all possible end actions σ, it follows that $\mathcal{E} \cap \mathcal{M}[X, \mu]$ is typical in $\mathcal{M}[X, \mu]$.

Suppose that (X, μ) has no ends of infinite measure. Then by property (ii) of Lemma 14.21 it follows that every homeomorphism in $\mathcal{M}[X, \mu]$ induces a zero charge on the ends. So as in the previous paragraph we have $\mathcal{M}_\sigma[X, \mu] = \mathcal{M}_\sigma^0[X, \mu]$ for all σ, and the result follows similarly from Theorem 15.1.

Conversely, suppose there are two ends e and e' of infinite measure. We sketch a proof that ergodicity cannot be dense in the compact-open topology on $\mathcal{M}[X, \mu]$. Suppose first that the manifold X is the n-dimensional tube $R^1 \times I^{n-1}$ and that the OU measure μ is infinite on both ends of the tube. We first define a volume preserving homeomorphism f of the tube $R^1 \times I^{n-1}$ which is an analog of the unit translation on the strip manifold which we considered in Example 13.2. The homeomorphism f on the tube moves points 'to the right by one unit' in the center of I^{n-1} and tapers off to the identity on the boundary of I^{n-1}. Then f is a volume preserving and end preserving homeomorphism of the tube with nonzero charge to the right. Because both ends of the tube have infinite μ-measure, the Homeomorphic Measures Theorem of Berlanga and Epstein (Theorem A2.8) implies that μ and volume measure λ are homeomorphic; i.e., there is an end preserving homeomorphism $h \in \mathcal{H}[X, \partial X]$ such that $\mu = \lambda h$. The homeomorphism $\hat{f} = h^{-1} f h$ is a μ-preserving end preserving homeomorphism which has nonzero charge and so cannot be ergodic by Theorem 14.23. If the manifold X is not the tube, we won't go into detail, but using Berlanga's structure theorem (Theorem A2.13), we can embed a tube from one end of infinite measure to the other. The analog of $\hat{f}$ to the embedded tube extended to all of X (by setting it to be the identity off the embedded tube) has nonzero charge on X, and so cannot be the compact-open limit of ergodic homeomorphisms (or even μ-recurrent homeomorphisms).   □

We have already seen that ergodicity $\mathcal{E}$ is not typical in the space of area preserving homeomorphisms of the strip $(\check{X}, \check{\mu})$ (in $\mathcal{M}[\check{X}, \check{\mu}]$). However, the question of whether or not there are *any* ergodic area preserving homeomorphisms of the strip can be answered by looking at just the *end reversing* (end transposing) homeomorphisms of $(\check{X}, \check{\mu})$ and asking whether ergodicity is typical in the component of $\mathcal{M}[\check{X}, \check{\mu}]$ which consists of just end reversing homeomorphisms. Indeed, the answer is yes and follows from a more general result. Specifically, for any induced end homeomorphism σ of the manifold (X, μ) we can ask whether ergodicity is typical in the component of μ-preserving homeomorphisms of X inducing the end homeomorphism σ, i.e., in $\mathcal{M}_\sigma[X, \mu]$. Furthermore we remind the reader that even if the action of σ at the ends is ergodic on $(E, \mathcal{Q}, \mu^*)$, σ need not be incompressible. (See Definition 14.11 and the examples after it.)

Corollary 15.4 *Suppose $\sigma = h^*$ for some $h \in \mathcal{M}[X, \mu]$. Then ergodicity is generic in $\mathcal{M}_\sigma[X, \mu]$ if and only if the dynamical system $(E, \mathcal{Q}, \mu^*, \sigma)$ induced on the ends is ergodic and incompressible.*

Proof Suppose that σ is ergodic and incompressible. It follows from ergodicity that for any σ-invariant set of ends I, either $\mu^*(I) = 0$ or $\mu^*(E - I) = 0$. In either case for any $g \in \mathcal{M}_\sigma[X, \mu]$, we have $0 = c_g(E) = c_g(I) + c_g(E - I)$. Part (ii) of Lemma 14.21 now shows that $c_g(I) = 0$ for every $I \in \mathcal{I}_\sigma$. Consequently c_g is identically zero on $\mathcal{I}_\sigma$. Thus $\mathcal{M}_\sigma^0[X, \mu] = \mathcal{M}_\sigma[X, \mu]$ and the result follows from Theorem 15.1 since σ is incompressible.

If σ is not incompressible then Lemma 14.15 states that $\mathcal{M}_\sigma[X, \mu]$ contains *no* μ-recurrent, and therefore no ergodic homeomorphisms. If σ is incompressible but not ergodic, then for some separating set K we can find a σ-invariant $I \in \mathcal{Q}_K$ with $\mu^*(I) \neq 0$ and $\mu^*(E - I) \neq 0$. Consequently we may choose infinite measured ends $e \in I$ and $e' \in E - I$. Then using a tube construction as in the proof of Corollary 15.3, construct an end preserving homeomorphism $\hat{f} \in \mathcal{M}[X, \mu]$ which 'flows an amount α of mass between the ends e and e''. For any $h \in \mathcal{M}_\sigma[X, \mu]$

the composition $\hat{f} \circ h$ belongs to $\mathcal{M}_\sigma[X, \mu]$ and has nonzero charge if α is chosen appropriately. Theorem 14.23 implies that an open neighborhood of this composition contains no ergodic homeomorphisms. $\qquad\square$

Of course we know that in order to have an ergodic homeomorphism with a specified induced end homeomorphism σ, it must be that σ is incompressible; furthermore if $\sigma = h^*$ for some $h \in \mathcal{M}[X, \mu]$ and the charge induced by h is not identically zero, then h cannot be ergodic. However, using a sequence of tube constructions as in the proof of the previous corollary, we can modify h to give another homeomorphism $\bar{h} \in \mathcal{M}_\sigma[X, \mu]$ whose charge $c_{\bar{h}}$ is identically 0. Thus, if $\mathcal{M}_\sigma[X, \mu]$ is not empty, then the set of zero-charge homeomorphisms with induced end homeomorphism σ, $\mathcal{M}_\sigma^0[X, \mu]$, is also not empty. Furthermore we can prove Baire category theorems in the latter space.

The next result now answers the question of when there *exist* ergodic homeomorphisms of (X, μ) with a specified end homeomorphism σ – this happens if and only if the end homeomorphism is incompressible.

Corollary 15.5 *Let $(E, \mathcal{Q}, \mu^*, \sigma)$ be the end dynamical system induced by some $h \in \mathcal{M}[X, \mu]$. Then $\mathcal{M}_\sigma[X, \mu]$ contains an ergodic homeomorphism if and only if $(E, \mathcal{Q}, \mu^*, \sigma)$ is incompressible.*

Proof The 'only if' part was already established in Lemma 14.15 where it was shown that if σ is compressible then $\mathcal{M}_\sigma[X, \mu]$ contains no recurrent homeomorphism.

Conversely suppose that $(E, \mathcal{Q}, \mu^*, \sigma)$ is incompressible. Recall that if $\mu(X) < \infty$ then σ is always incompressible and furthermore $\mathcal{M}_\sigma[X, \mu] = \mathcal{M}_\sigma^0[X, \mu]$ (see Example (iii) following Lemma 14.21); consequently Theorem 15.1 implies that the ergodic homeomorphisms are dense in $\mathcal{M}_\sigma[X, \mu]$. If however $\mu(X) = \infty$, then the comment preceding this corollary shows that $\mathcal{M}_\sigma^0[X, \mu]$ is not empty, and again Theorem 15.1 guarantees ergodic homeomorphisms are generic in $\mathcal{M}_\sigma^0[X, \mu]$. $\qquad\square$

We now consider the question of which measured manifolds (X, μ) support *some* ergodic μ-preserving homeomorphism. Observe that the identity homeomorphism $i \in \mathcal{M}[X, \mu]$ induces the identity homeomorphism i^* on the ends of X, which is obviously incompressible. So the previous result shows that $\mathcal{M}_{i^*}[X, \mu]$ contains an ergodic homeomorphism. So the answer to the original question is that *any* measured manifold supports an ergodic measure preserving homeomorphism, or more precisely,

Theorem 15.6 *Let μ be an OU measure on a sigma compact manifold X. Then there exists an ergodic, μ-preserving homeomorphism of X.*

The next result asserts that when the action at the ends is transitive then ergodicity is typical in the component of μ-preserving homeomorphisms inducing that end homeomorphism.

Corollary 15.7 *Ergodicity is generic in $\mathcal{M}_\sigma[X,\mu]$ if the restriction of σ to E^∞, the ends of infinite measure, is transitive.*

Proof For such a σ, the system $(E, \mathcal{Q}, \mu^*, \sigma)$ is ergodic and incompressible, so the result follows from Corollary 15.4. $\qquad\square$

We note that when σ is transitive on E, either $\mu(X) < \infty$, or all of the ends have infinite measure. In either case, Theorem 15.2 or the previous corollary implies ergodicity is a compact-open dense G_δ subset of $\mathcal{M}_\sigma[X,\mu]$. We state this as follows.

Corollary 15.8 *If σ is transitive on E then ergodicity is generic in $\mathcal{M}_\sigma[X,\mu]$.*

Combining Theorem 15.1, Lemma 14.15, and Theorem 14.23, we obtain the following characterization of the closure of the ergodic μ-preserving homeomorphisms of a manifold, a result we described in Section 13.3 as Theorem F.

Corollary 15.9 *A μ-preserving homeomorphism h of (X,μ) is in the compact-open closure of the ergodic homeomorphisms if and only if h induces an incompressible homeomorphism of the ends of X and induces an identically zero charge c_h on these ends.*

Proof If h^* is incompressible and $c_h \equiv 0$, then Theorem 15.1 ensures that ergodicity is a dense property in $\mathcal{M}_{h^*}^0[X,\mu]$. Conversely if h^* is compressible then no homeomorphism in its clopen neighborhood $\mathcal{M}_{h^*}[X,\mu]$ can be ergodic, by Lemma 14.15. Similarly if c_h is not identically zero, the result follows from Theorem 14.23. $\qquad\square$

16

Ergodic Homeomorphisms: Proofs

16.1 Introduction

In the previous chapter we presented our main group of results concerning ergodic homeomorphisms of noncompact manifolds. We gave necessary and sufficient conditions for ergodicity to be generic in the whole space $\mathcal{M}[X,\mu]$, and in a given component $\mathcal{M}_\sigma[X,\mu]$. We also gave some existence results for ergodic μ-preserving homeomorphisms with certain end related properties and a necessary and sufficient condition that a μ-preserving homeomorphism lies in the compact-open topology closure of ergodic homeomorphisms. All these results are related and indeed we showed how they all follow from the result on genericity of ergodicity in $\mathcal{M}_\sigma^0[X,\mu]$, which we restate here.

Theorem 15.1 *If $(E, \mathcal{Q}, \mu^*, \sigma)$ is incompressible then the ergodic homeomorphisms of the sigma compact connected manifold (X,μ) with end action σ and identically zero end charge, i.e., $\mathcal{E} \cap \mathcal{M}_\sigma^0[X,\mu]$, form a compact-open dense G_δ subset of $\mathcal{M}_\sigma^0[X,\mu]$, the set of μ-preserving homeomorphisms with induced end homeomorphism σ and identically zero end charge.*

This chapter is mainly devoted to proving Theorem 15.1. The modifications needed to prove Theorem 15.2 (for $\mu(X) < \infty$) are given at the end of the chapter. The proof is based partly on ideas used earlier in the first proof of generic ergodicity for the cube (Theorem 7.1), so the reader is urged to review that proof. In particular, both proofs show that a certain approximating automorphism is ergodic by showing that it can be viewed as a skyscraper built over an ergodic base. The well known fact that such an automorphism is ergodic is presented in Theorem A1.1. The first appendix gives a review of skyscraper constructions, so the reader unfamiliar with these should read that appendix

137

before attempting to follow the proof of Theorem 15.1. That appendix also contains a conjugacy approximation result (Corollary A1.12) which will be needed in the proof.

This chapter is organized as follows. In Section 16.2 we give an informal outline of the proof (of Theorem 15.1). Since the complexities of the proof are due to problems in end dynamics, we present in Section 16.3 a proof of Theorem 15.1 for the special case of end preserving homeomorphisms of the strip manifold. Since this manifold has only two ends, certain problems in the general case are avoided, while still giving the reader an idea of how problems with ends can be handled. In Section 16.4 we begin our proof of Theorem 15.1 for the general case by presenting a skyscraper construction which produces an approximating ergodic *automorphism*, which however is not in general a homeomorphism. Section 16.5 completes the proof of Theorem 15.1 by showing how the End Preserving Lusin Theorem, Theorem 14.9, can be used to find an approximating ergodic *homeomorphism*.

The material presented in this chapter was first obtained [20] for manifolds like the strip manifold which have only finitely many ends, using techniques similar to those of Section 16.3. The generalization to arbitrary sigma compact manifolds came the following year [22]. Both results rely heavily on the conjugacy results which follow from the Multiple Tower Rokhlin Theorem [13], which is proved in Appendix 1.

16.2 Outline of Proofs of Theorems 15.1 and 15.2

Since our proof of Theorem 15.1 is somewhat complicated, we provide the reader with this brief section which gives a broad outline of the main structure of the proof. This outline presents the elements of the proof in reverse order, so that the earlier constructions are motivated by the requirements of the later stages. Much of this outline will be repeated in the actual proofs.

As with all of our genericity proofs (except for the 'classical proofs' given in Sections 7.2 and 12.3), we begin by writing the set we hope is generic in $\mathcal{M}[X,\mu]$ as the countable intersection of weak topology open subsets of $\mathcal{G}[X,\mu]$. So here we write the ergodic automorphisms as $\mathcal{E} = \bigcap_{i=1}^{\infty} \mathcal{E}_i$, where each set $\mathcal{E}_i$ is open in $\mathcal{G}[X,\mu]$ in the weak topology. We need to show that each set $\mathcal{E}_i$ is dense in $\mathcal{M}_\sigma^0 = \mathcal{M}_\sigma^0[X,\mu]$ in the compact-open topology, from which it will follow by the Baire Category Theorem that $\mathcal{E} \cap \mathcal{M}_\sigma^0$ is a dense G_δ subset of $\mathcal{M}_\sigma^0$. So writing an arbitrary set $\mathcal{E}_i$ as $\mathcal{F}$, we need to show that the 'target' set $\mathcal{T}$ defined by

$$\mathcal{T} = \mathcal{F} \cap \mathcal{M}_\sigma^0 \cap \mathcal{C} \text{ is not empty}, \tag{16.1}$$

where $\mathcal{C} = \mathcal{C}(h, K, \epsilon)$ is an arbitrary compact-open basic open neighborhood of a given homeomorphism $h \in \mathcal{M}$. By Lemma 14.26 we may assume that the compact set K is h-moving. This means that none of the sets K or $P(K)$, for P in $\mathcal{P}_K$, are h-invariant. This is essential to the construction as the ergodic approximation to h will agree with h setwise on K and each set $P(K)$. The combined constructions in the proof of Theorem 15.1 produce a small homeomorphism h' with hh' belonging to the target set $\mathcal{T}$, which shows that $\mathcal{T}$ cannot be empty.

The proof begins with a skyscraper construction with an ergodic base transformation which gives an ergodic (by Theorem A1.2) *automorphism* f approximating h in the sense that f belongs to the neighborhood $\mathcal{C}$. We then approximate the 'difference' $g = h^{-1}f \in h^{-1}\mathcal{F}$ by an end preserving *homeomorphism* $h' \in \mathcal{M}_{id}^0$, using the End Preserving Lusin Theorem, Theorem 14.9. Referring to the hypotheses of this theorem, we see that we require that the automorphism $g = h^{-1}f$ must leave invariant the compact set K as well as each of the components $P(K)$, $P \in \mathcal{P}_K$, of its complement. This is the same as requiring that the ergodic approximation f agrees (setwise) with the given homeomorphism h on each of these sets. So when we construct the ergodic automorphism f we need to ensure not only that it belongs to the compact-open neighborhood $\mathcal{C}$, but also that it agrees setwise with h on the partition determined by K and the components of its complement.

The construction of the ergodic automorphism approximation f to h is carried out in Lemma 16.3. This lemma constitutes the heart of the proof. In the special case that the manifold has finite measure, a different construction of the approximation f is carried out in Lemma 16.2. This lemma has a stronger conclusion than its infinite measure counterpart in that the automorphism f it produces is not only ergodic, but moreover can be made to be conjugate to any given antiperiodic automorphism of a finite Lebesgue space. For example, it could produce a weak mixing automorphism f if desired. The reason for the different conclusions of the two lemmas is the nature of the constructions they employ. In the infinite measure version, Lemma 16.3, the automorphism f is obtained by a skyscraper construction with an ergodic base. In the finite measure version, Lemma 16.2, the automorphism f is obtained using a general conjugacy result called the Setwise Conjugacy Approximation Theorem, which is proved in Appendix 1 as Corollary A1.12 (of the Multiple Tower Rokhlin Theorem).

In both the finite and infinite measure versions, the approximation of the difference automorphism $g = h^{-1}f$ by a measure preserving homeomorphism h' is carried out using the End Preserving Lusin Theorem 14.9. The homeomorphism hh' will belong to the required target set $\mathcal{T}$ in equation (16.1) (hh' will be an 'approximately ergodic' homeomorphism near our given h). An appeal to the Baire Category Theorem will complete the proof of the main theorems (15.1 and 15.2).

The approximation result referred to above is Corollary A1.12, which we restate here in a notation which will cause less confusion than the original would in our applications.

Corollary A1.12 (Setwise Conjugacy Approximation) *Denote by* (Y, Σ, μ), *the finite measure space* Y $(\mu(Y) < \infty)$, *with sigma algebra* Σ *on which the measure* μ *is defined. Let* $\theta_1, \theta_2 \in \mathcal{G}[Y, \mu]$, *with* θ_2 *anti-periodic. Let* Λ *be a finite subalgebra of* Σ *such that* $\theta_1|\Lambda$, *the restriction of* θ_1 *to the algebra* Λ, *has no nontrivial periodic set. Further, let d be any metric on* Y *such that* Y *is totally bounded and all nonempty open sets have positive measure. Let* C *denote the union of all atoms of* Λ *whose image under* θ_1 *is connected and relatively compact. Then given any* $\epsilon > 0$, *there is a conjugate* $\hat{\theta}_2 = k^{-1}\theta_2 k$, $k \in \mathcal{G}[Y, \mu]$, *of* θ_2 *such that* $\hat{\theta}_2(A) = \theta_1(A)$, $\forall A \in \Lambda$, *and* $d(\hat{\theta}_2(x), \theta_1(x)) < \epsilon$ *for all* $x \in C$.

Note that this is an extension of Theorem 10.1 in which the conjugacy approximation (here called $\hat{\theta}_2$) is not only close to its target (θ_1) but also agrees setwise with it on given sets.

This outline describes the proof of the general results, Theorems 15.1 and 15.2. However, the ideas apply equally to the strip manifold and in the next section we use this special case as a useful introduction to the ideas used in the proof of Theorem 15.1.

16.3 Proof of Theorem 15.1: Strip Manifold

To illustrate the ideas in the full proof of Theorem 15.1, we first give a proof for the special case of zero charge, end preserving, area preserving homeomorphisms of the strip manifold, i.e., for $\mathcal{M}_{id}^0[R^1 \times [0,1], \lambda]$. This section can be skipped, as we will not use the partial result to obtain the more general case.

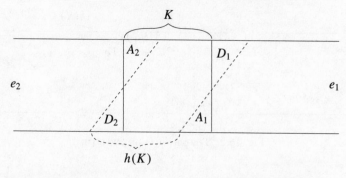

Fig. 16.1. The arrival sets A_i and departure sets D_i on the strip

Theorem 16.1 *The ergodic end preserving homeomorphisms form a compact-open topology dense G_δ subset of the zero charge, end preserving, area preserving homeomorphisms of the strip $R^1 \times [0,1]$.*

Proof Let $h \in \mathcal{M}_{id}^0[R^1 \times I, \lambda]$ be a zero charge, end preserving, area preserving homeomorphism of the strip, and $\mathcal{C}(h, K, \epsilon)$ a compact-open topology basic neighborhood of h. The first step is to approximate h by an ergodic *automorphism* f in $\mathcal{C}(h, K, \epsilon)$. In order to do this, we first make some assumptions on h and K: namely, by Lemma 14.26, there is no loss of generality in assuming that $K = [a, b] \times [0, 1]$ and is h-moving. The strip has two ends: let e_1 be the right hand end and e_2 the left hand end. For $i = 1, 2$, define $A_i = K \cap h(e_i(K))$, and let $D_i = e_i(K) \cap h(K)$. We note that $A = A_1 \cup A_2$ is the 'arrival set' into K and $D = D_1 \cup D_2$ is the 'departure set' from K. This means that when an h-orbit enters K it enters via A and when it leaves K it enters D. The assumption that h has zero charge means that $\lambda(A_i) = \lambda(D_i)$ for each $i = 1, 2$. We picture these sets and the action of h on the strip in Figure 16.1.

In Figure 16.1 we have chosen K sufficiently large so that $h(e_i(K)) \cap e_j(K) = \emptyset$ for $i \neq j$. Set $J_{ij} = e_i(K) \cap h^{-1}(e_j(K))$ and

$$J = \bigcup_{ij}(e_i(K) \cap h^{-1}(e_j(K))) = \bigcup_{ij} J_{ij}$$

because in general the figure might look like that in Figure 16.2, where J is not empty and its image $h(J)$ is shown.

The construction of the ergodic automorphism f which approximates h is done in two steps. First we use a finite measure technique (namely the Setwise Conjugacy Approximation) to construct an *ergodic*, area

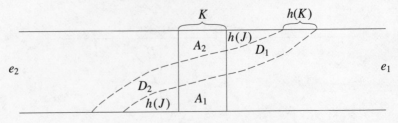

Fig. 16.2. The set $h(J)$ on the strip

preserving automorphism of the finite measured set $Y = K \cup h(K) \cup J \cup h(J)$ onto itself. (Note that $K - h(K) = A$ and $h(K) - K = D$.)

We would like to apply the conjugacy approximation (Corollary A1.12) stated at the end of the previous section to h – unfortunately h does not map $Y = K \cup h(K) \cup J \cup h(J)$ to itself. Thus we replace h by an *automorphism* $\tilde{h}$ of Y which agrees with h on $K \cup J$ and maps D_i onto A_i for each $i = 1, 2$. This is possible precisely because the zero charge assumption has given the equalities $\lambda(A_i) = \lambda(D_i)$, $i = 1, 2$. Now apply the Setwise Conjugacy Approximation on Y to the automorphism $\theta_1 = \tilde{h}^{-1}$, with the second automorphism θ_2 any ergodic area preserving automorphism of the finite measured set Y. Let Λ be the algebra whose atoms are $h(K), h(J_{ij}), A_1$, and A_2. The condition of the Setwise Conjugacy Approximation that $\tilde{h}^{-1}$ restricted to Λ has no periodic set follows from the assumption that K is h-moving. Since $\theta_1(h(K)) = \tilde{h}^{-1}(h(K)) = K$, which is compact and connected, we have $C = h(K)$. If we take $\tilde{f}^{-1} : Y \to Y$ to be the conjugate of θ_2 given by the Setwise Conjugacy Approximation, then $\tilde{f}$ is an ergodic area preserving automorphism of the finite measured set $Y = K \cup h(K)$ satisfying

(i) $d(\tilde{f}^{-1}(y), h^{-1}(y)) < \delta$ for a.e. y in $h(K)$
(ii) $\tilde{f}(K) = h(K)$
(iii) $\tilde{f}(D_i) = A_i$, for $i = 1, 2$.

The second step is to extend the restriction of $\tilde{f}$ to $Y - (D_1 \cup D_2)$ to an ergodic automorphism f of the entire strip which is setwise the same as h on the components $e_i(K)$ for $i = 1, 2$, i.e., satisfying

(i) $d(f^{-1}(y), h^{-1}(y)) < \delta$ for a.e. y in $h(K)$
(ii) $f(K) = h(K)$
(iii) $f(e_i(K)) = h(e_i(K))$, for $i = 1, 2$.

This is accomplished using 'skyscraper constructions' (see Appendix 1). We can picture the action of $\tilde{f}$ on Y as a skyscraper over the set A

and let $\bar{f}$ be the first return map on A. The transformation $\tilde{f}$ can be represented as a skyscraper over the set A as the base, D as the union of the top levels of the skyscraper and $\bar{f}$ as the base transformation. Since $\tilde{f}$ is ergodic on Y, it follows from Theorem A1.1, that $\bar{f}$ is ergodic on A. The construction of f will be completed by stacking additional levels on top of the skyscraper using sets from the complement of Y. Furthermore we ensure that all points in the complement of Y appear somewhere in this skyscraper. Once this is done, the required automorphism in $\mathcal{G}[R^1 \times I, \lambda]$ is defined as follows. All points not on a top level are mapped by f onto the point directly above (one level up). For a point x lying on a top level of the skyscraper, define $f(x) = \bar{f}(y)$ where y is the point on the base A lying directly below x. The ergodicity of $\bar{f}$ will guarantee that the resulting automorphism f defined on the strip is ergodic and that any point leaving Y by D_i eventually returns to Y through A_i (for each $i = 1, 2$). So if we identify all levels stacked above an A_i with points in $e_i(K)$ for each $i = 1, 2$, the automorphism f will satisfy $f(e_i(K)) = h(e_i(K))$. This completes the second step of the proof.

Observe that $g = h^{-1}f$ is an automorphism which belongs to the weak topology open set $h^{-1}\mathcal{F}$ (since f is ergodic and $\mathcal{F}$ contains the ergodics), is uniformly small on the compact set K and leaves invariant the two components $e_1(K)$ and $e_2(K)$. Indeed, all of the levels of the skyscraper were added in the order that they appear purely to guarantee the same setwise behavior for f and h. Consequently $g = h^{-1}f$ leaves invariant the sets $e_i(K)$. The End Preserving Lusin Theorem 14.9 applied to $g = h^{-1}f$ says that there is an area preserving homeomorphism h' of the strip with these same properties: $h' \in h^{-1}\mathcal{F}$, h' is small on K, and leaves invariant K and $e_i(K)$, $i = 1, 2$. These properties imply that $hh' \in \mathcal{F} \cap \mathcal{M}_{id}^0[R^1 \times I, \lambda] \cap \mathcal{C}(h, K, \epsilon)$. Consequently we have shown the target set defined in equation (16.1) is not empty, and the result follows by the Baire category argument given before that equation. $\qquad\square$

16.4 Proofs of Theorems 15.1 and 15.2: General Case

In the general case of the proof of Theorem 15.1 the basic strategy is the same as for the strip manifold, but the details are more complicated because there may be ends of finite measure which must be treated separately. Furthermore the dynamics on the ends may be more complicated and consequently much more care needs to be taken when doing the skyscraper construction.

We proceed with the proof of the main theorem. The proof of Theorem 15.1 begins by approximating the measure preserving homeomorphism h by an ergodic automorphism f in the following pointwise–setwise manner:

Any μ-preserving homeomorphism h of X which induces an incompressible end homeomorphism σ with zero end charge can be approximated by an ergodic μ-preserving automorphism f in the following manner: f is pointwise close on a given h-moving compact set K, and setwise the same on all sets $P(K)$ where $P \in \mathcal{P}_K$.

This result will be proved proved in two different settings, depending on whether or not $\mu(X)$ is finite. When $\mu(X)$ is finite, *every μ-preserving* homeomorphism h induces an incompressible end homeomorphism and has zero end charge. Both of these settings for the approximation above need the Setwise Conjugacy Approximation Theorem stated earlier (this is Corollary A1.12).

This corollary is applied to prove the pointwise–setwise approximation to h by an ergodic automorphism f in the finite measure case – again as before, for technical reasons we will apply the Setwise Conjugacy Theorem to h^{-1} rather than h. We record this finite measure case approximation as the following lemma.

Lemma 16.2 *Let μ be a finite OU measure on a sigma compact manifold X. Let h be a μ-preserving homeomorphism of X and let K be any (separating) h-moving compact subset of X. Set $K = X$ if X is compact. Then for any antiperiodic μ-preserving automorphism θ of X, and any $\delta > 0$, there is an automorphism γ in $\mathcal{G}[X, \mu]$ such that the conjugate f of θ given by $f = \gamma^{-1}\theta\gamma$ satisfies*

(i) $d(f^{-1}(y), h^{-1}(y)) < \delta$ *for μ-a.e. $y \in h(K)$*

(ii) $f(K) = h(K)$

(iii) $f(P(K)) = h(P(K))$ *for every set of ends $P \in \mathcal{P}_K$.*

Proof First suppose that X is not compact. Apply the Setwise Conjugacy Approximation Theorem (Corollary A1.12) in the formulation given at the end of Section 16.2 to $Y = X$, $\theta_1 = h^{-1}$, and $\theta_2 = \theta^{-1}$, with Λ the algebra consisting of sets of the form $hQ(K)$ or $hQ(K) \cup hK$ where Q varies over $\mathcal{Q}_K$. The assumption that K is h-moving ensures that the automorphism $(h^{-1})|\Lambda$ of the algebra Λ, has no nontrivial periodic set. The set C, which is the union of all the atoms of Λ whose image under h^{-1} is connected and relatively compact, is just the single atom $h(K)$. The automorphism $f \in \mathcal{G}[X,\mu]$ is defined by $f^{-1} = \hat{\theta}$, where $\hat{\theta}$ is the conjugate of θ guaranteed by the Setwise Conjugacy Approximation Theorem. This f satisfies the requirements of the theorem: since if $f^{-1}(h(K)) = h^{-1}(h(K))$, then $f(K) = h(K)$; the third property that $f(P(K)) = h(P(K))$ follows similarly. In the case that $X = K$ is compact, the above proof works with the trivial algebra $\Lambda = \{X, \emptyset\}$ (or actually it is the first part of the Setwise Conjugacy Approximation Theorem). $\qquad\square$

Note that in the proof above the Setwise Conjugacy Approximation Theorem (Corollary A1.12) was applied to $Y = X$ since X had finite μ-measure. Furthermore Lemma 16.2 proves that f is conjugate to *any given antiperiodic automorphism* $\theta \in \mathcal{G}[X,\mu]$. Consequently, if we choose θ to be any ergodic automorphism of (X,μ), we can approximate the homeomorphism by an ergodic automorphism f in the pointwise–setwise fashion when $\mu(X) < \infty$. But Lemma 16.2 proves more. Let $\mathcal{F}$ be any conjugate invariant subset of $\mathcal{G}[X,\mu]$. Note that the conjugate (by any automorphism in $\mathcal{G}[X,\mu]$) of any antiperiodic automorphism $\theta \in \mathcal{F}$ also belongs to $\mathcal{F}$. Consequently, by choosing $\mathcal{F}$ to be (for example) the weak mixing automorphisms, or the zero entropy automorphisms of (X,μ), we can require that the approximation f is also weak mixing or has zero entropy.

In the next theorem, which is the same approximation for the case when the OU measure μ is infinite, the Setwise Conjugacy Approximation is applied to only a finite measure part of the space and we do not assert that the resulting conjugate has any properties other than ergodicity. The theorem uses Corollary A1.12 and also relies extensively on skyscraper constructions from ergodic theory (see Appendix 1 and Friedman's book [63]).

Lemma 16.3 *Assume μ is an infinite OU measure on a sigma compact connected manifold X. Let h be a μ-preserving homeomorphism of X which induces an incompressible end homeomorphism $\sigma = h^*$ with an identically zero charge on the ends E. Then given any h-moving set K in X, and any $\delta > 0$, there is an ergodic μ-preserving automorphism $f \in \mathcal{G}[X,\mu]$ such that*

(i) $d(f^{-1}(y), h^{-1}(y)) < \delta$ *for μ-a.e. $y \in h(K)$*

(ii) $f(K) = h(K)$

(iii) $f(P(K)) = h(P(K))$ *for every set of ends* $P \in \mathcal{P}_K$.

Proof In order to obtain an approximant f to h which is setwise the same as h on the components of $X - K$, we need to look at the behavior of h on these components a little more closely than in the previous lemma for the finite measure case (where all of the complications were dealt with by the Setwise Conjugacy Approximation Theorem, Corollary A1.12). Let $\mathcal{P}_K = \{E_1, \ldots, E_m, E_{m+1}, \ldots, E_l\}$ be the partition of E associated with the compact set K. The connected components of $X - K$ are $E_1(K), \ldots, E_m(K), E_{m+1}(K), \ldots, E_l(K)$ and have been labeled so that the $E_i(K)$ have infinite μ-measure for $i \leq m$ and finite μ-measure for $i > m$. For $i, j = 1, \ldots, l$ define $J_{ij} = J_{ij}(K) = E_i(K) \cap h^{-1}E_j(K)$ and $t_{ij} = 1$ if $\mu(J_{ij}) = \infty$ and 0 otherwise. Note that if i or j exceeds m then $t_{ij} = 0$. Define

$$J = \bigcup_{\{i,j : t_{ij} = 0\}} J_{ij}$$

and observe that $\mu(J) < \infty$. The set $h(J)$ is pictured in Figure 16.2 for the strip manifold. Given our assumption of incompressibility, Lemma 14.16 implies that the state space $\{1, \ldots, m\}$ can be partitioned into communicating states so that the associated submatrices of $\mathbf{T} = (t_{ij})$ are irreducible. Suppose there are p such sets. Thus we have a relabeling of the clopen sets of ends $E_1, \ldots, E_m$ as C_i^r where $i = 1, \ldots, p$, and $r = 1, \ldots, q_i$. For each $i = 1, \ldots, p$, we let $C_i = \bigcup_{r=1}^{q_i} C_i^r$ denote the ith class of communicating states (ends), and note that C_i is an h^*-invariant set of ends (i.e., $C_i \in \mathcal{I}_h$). Note that for the strip manifold case in the previous section, we have $E_1 = \{e_1\} = C_1$ and $E_2 = \{e_2\} = C_2$ and $\mathbf{T}$ is the 2×2 identity matrix. In Example 14.17 (using the notation of the four-ended cylinder) we have $C_1 = \{e_{-\infty}\}$ and $C_2 = Q_1 \cup Q_2 = \{e_1, e_2, e_3\}$.

Thus the components of $X - K$ are grouped into the infinite measured sets $C_i(K)$ for $i = 1, \ldots, p$ and the finite measured components $\bigcup_{i=m+1}^l E_i(K)$. For $i = 1, \ldots, p$, define the 'arrival set from $C_i(K)$' by $A_i = h(C_i(K)) - C_i(K)$ and let $D_i = C_i(K) - h(C_i(K))$ (this is the set of points in X which have 'departed into $C_i(K)$'). Our assumption that the end charge induced by h is identically zero implies $0 = c_h(C_i) = c_h(C_i, K) = \mu(D_i) - \mu(A_i)$. Consequently for each $i = 1, \ldots, p$, we have that $\mu(A_i) = \mu(D_i) < \infty$, and the assumption that K is h-moving implies that for each i, this common number is positive. Denote by $A = \bigcup_{i=1}^p A_i$ the 'arrival set from the infinite measured components' and by $D = \bigcup_{i=1}^{p} D_i$ the 'departure set into the infinite measured components'. Note that $\mu(A) = \mu(D)$ and that both of these sets have finite and positive μ-measure.

Let Y be the finite measured set given by

$$Y = K \cup hK \cup J \cup h(J).$$

We would like to apply the Setwise Conjugacy Approximation Theorem to h on Y, but h does not map Y onto itself. Consequently, we first replace h by *an automorphism* $\tilde{h}$ of Y which agrees with h on $K \cup J$. Define $\tilde{h} : Y \to Y$ to be any μ-preserving automorphism of Y which agrees with h on $K \cup J$ and maps D_i onto A_i for $i = 1, \ldots, p$. This is possible because $\mu(D_i) = \mu(A_i)$ for $i = 1, \ldots, p$. Let us compare h and $\tilde{h}$. Suppose $x \in X$ is a point whose h-orbit enters Y, with

$$h^{a-1}(x) \notin Y, \; h^a(x) \in Y, \; h^{a+1}(x) \in Y, \ldots, h^d(x) \in Y, \; h^{d+1}(x) \notin Y.$$

Then $h^a(x) \in A$, $h^d(x) \in D$ and $\tilde{h}(y) = h(y)$ for $y = h^i(x)$, $a < i < d$.

Now apply the Setwise Conjugacy Approximation Theorem (Corollary A1.12 as restated in the notation earlier in this chapter) on Y to the automorphism $\theta_1 = \tilde{h}^{-1}$, with the second automorphism $\theta_2 \in \mathcal{G}[Y, \mu]$ being any ergodic automorphism of Y. The algebra Λ is the one whose atoms are $h(K)$, $h(J_{ij})$ for i, j with $t_{ij} = 0$, and A_i for $i = 1, \ldots, p$. The condition that $\tilde{h}^{-1}$ restricted to Λ has no nontrivial periodic set follows from the assumption that K is h-moving. The set C in this application of Corollary A1.12 is again $h(K)$. If we take $\tilde{f}^{-1} : Y \to Y$ to be the conjugate of θ_2 given by Corollary A1.12 then $\tilde{f}$ is an ergodic μ-preserving automorphism of the finite measured set Y satisfying

(i) $d(\tilde{f}^{-1}(y), h^{-1}(y)) < \delta$ for μ-a.e. y in $h(K)$

(ii) $\tilde{f}(K) = h(K)$

(iii) $\tilde{f}(J_{ij}) = h(J_{ij})$, for i, j with $t_{ij} = 0$ and $\tilde{f}(D_i) = A_i$, $i = 1, \ldots, p$.

The next step in the the construction of the automorphism f of the whole space, the ergodic approximation to h, is to 'extend' $\tilde{f}$ from an ergodic automorphism of Y to an ergodic automorphism of X (note that we use the term 'extend' loosely – technically f extends the restriction of $\tilde{f} : Y - D \to Y - A$). To do this we represent the ergodic automorphism $\tilde{f} : Y \to Y$ as a skyscraper over A. Note that this description of skyscraper constructions is also outlined in more detail in the first Appendix. By this we mean the following (using the terminology from skyscrapers described in Appendix 1): Let $\bar{f} : A \to A$ be the first return map of $\tilde{f}$ to the subset $A \subset Y$; i.e., for each $x \in A$ we define

$$\bar{f}(x) = \tilde{f}^{r_A(x)}(x)$$

where $r_A(x)$ is the smallest positive integer $r(x)$ such that $\tilde{f}^{r(x)}(x) \in A$. Since $\tilde{f}$ is ergodic on (Y, μ), it follows from Theorem A1.1 that $\bar{f}$ is ergodic on (A, μ). The set A can be partitioned into sets $A^k = \{x \in A : r_A(x) = k\}$, $k = 1, 2, \ldots$. Then the skyscraper over A is the triangular partition of Y given by the disjoint union

$$Y = \bigcup_{k=1}^{\infty} \bigcup_{i=1}^{k} \tilde{f}^{i-1} A^k.$$

The set $\bigcup_{i=1}^{k} \tilde{f}^{i-1} A^k$ is called the *column of height k* (over A), and A^k its *base* and $\tilde{f}^{k-1} A^k \subset D$ its *top*. When drawn in a diagram such as Figures 16.3 and A1.3, the action of $\tilde{f}$ on any y in the column of height k that is not in the top of the column is to map it to the point in the column directly above it in the orbit of $\tilde{f}$. A point in the top of the column, say $y = \tilde{f}^{k-1}(x)$ for some $x \in A^k$, is mapped to the point $\bar{f}(x)$ in A, the base of the skyscraper. This representation of Y and $\tilde{f}$ and its action in terms of the columns over A and the induced transformation $\bar{f}$ is called the skyscraper in Y over A.

In this final part of the construction we place additional levels over the skyscraper and fill these levels with all the points of $X - Y$. The required transformation $f : X \to X$ will be the resulting skyscraper in X over A. Theorem A1.2 now implies that f is ergodic because the first return map for f to the set A is $\bar{f}$, an ergodic automorphism of A. Note that in our construction of $\tilde{f}$, any point leaving Y by D_i will return to Y in A_i. So if we identify all the points in $X - Y$ stacked above D_i with points in $\hat{C}_i = C_i(K)$, the resulting automorphism will satisfy not quite condition (iii) of the lemma, but at least the weaker condition

$$f(C_i(K)) = h(C_i(K)) \text{ for all } i = 1, \ldots, p.$$

To ensure property (iii) we proceed as follows. Fix some index i, $i = 1, \ldots, p$. For the rest of this argument i will be suppressed except occasionally for emphasis. Consequently the ends C_i^r for $r = 1, \ldots, q$ ($= q_i$) in the ith class of communicating ends will be denoted simply by C^r. Let $\mathbf{B}$ denote the $q \times q$, 0–1 irreducible submatrix of the $m \times m$ matrix $\mathbf{T}$ corresponding to this irreducible set of q communicating states (see also Example 14.17, where $\mathbf{T}$ and $\mathbf{B}$ are given explicitly for a homeomorphism of the four-ended cylinder).

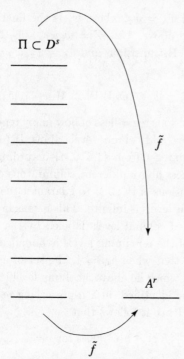

$$\Pi \subset D^s$$

$$\tilde{f}$$

$$A^r$$

$$\tilde{f}$$

Fig. 16.3. A subcolumn whose top Π is a subset of D^s and enters A^r under $\tilde{f}$

Since **B** is irreducible we can find a finite word $W = w_1 w_2 \ldots w_z$ in the symbols $\{1, \ldots, q\}$ such that all transitions $w_j w_{j+1}$ are 'legal' (i.e., $b_{w_j w_{j+1}} = 1$) and all legal transitions appear in W (if $b_{kr} = 1$, then for some j, $w_j = k$ and $w_{j+1} = r$). Furthermore we can ensure that the word W is cyclic in the sense that $w_z w_1$ is a 'legal' transition. (For an explicit construction of such a word W see the Example 16.4 at the end of this proof.)

Observe that $A_i = h(C_i(K)) - C_i(K)$ is the disjoint union of the sets $A^r = h(C_i^r(K)) \cap C_i^r(K)$, $r = 1, \ldots, q$ and $D_i = C_i(K) - h(C_i(K))$ is the disjoint union of the sets $D^s = C_i^s(K) \cap h(C_i^s(K))$, $s = 1, \ldots, q$. (Recall that our notation for A^r and D^s suppresses the dependence on the fixed index i for simplicity.)

Now fix any column of the skyscraper picture for $\tilde{f} : Y \to Y$ with base A. Consider a subcolumn whose top Π is a subset of D^s and such that $\tilde{f}(\Pi) \subset A^r$. See Figure 16.3.

Let $\hat{W} = \hat{w}_0 \hat{w}_1 \ldots \hat{w}_\alpha$ be a word in $\{1, \ldots, q\}$ with legal **B**-transitions,

$\alpha \leq q$, $\hat{w}_0 = s$, and $\hat{w}_\alpha = w_z$, where w_z is the final letter (number) of the word W defined above. Let $\check{W} = \check{w}_0\check{w}_1\ldots\check{w}_\beta$ be another word in $\{1,\ldots,q\}$ with legal **B**-transitions and $\check{w}_0 = w_z$, $\check{w}_\beta = r$, and $2 \leq \beta \leq q$. Then the word

$$W^* = \hat{w}_1\ldots\hat{w}_\alpha WW\ldots W\check{w}_1\ldots\check{w}_{\beta-1}$$

has all legal **B**-transitions regardless of how many repetitions of the word W it contains. So we add column levels above Π (of width $\mu(\Pi)$) and label each with an integer from $\{1,\ldots,n\}$ according to the word W^*. We repeat this process above different column tops of type Π, choosing the number of repetitions of W so as to guarantee that the total measure of the added columns levels is infinite. This means in particular that the measure of the added column levels labeled by the word W is infinite since the measure of the remaining levels is bounded by $2q\mu(D_i)$ (since $\alpha, \beta \leq q$). Observe that when $b_{sr} = 1$, the set of added column levels whose label is s and which lie above a column level labeled r has infinite measure and may be identified, in a μ-preserving manner, with the set $C_i^s(K) \cap hC_i^r(K)$. Hence it follows that

$$f(C_i^r(K)) = h(C_i^r(K)) \text{ for } r = 1,\ldots,q,$$

so that condition (iii) is satisfied.

$\square$

Example 16.4 *To illustrate the labeling in the previous proof we give the following example. Consider the matrix* **B** *on three states* $1, 2, 3$:

$$\mathbf{B} = \begin{pmatrix} 0 & 1 & 1 \\ 1 & 1 & 0 \\ 1 & 0 & 0 \end{pmatrix}.$$

The word $W = w_1w_2\ldots w_z = 12213$ *with* $z = 5$ *has only legal transitions and the legal transition from 3 to 1 appears as* w_5w_1 *which also makes* W *cyclic.*

In the above proof if we let $r = 2$ *and* $s = 3$, *then we need the word* W^* *to connect the state 2 to the state 3; i.e.,* $2W^*3$ *should have only legal* **B**-*transitions. The word* $\hat{W} = \hat{w}_0\hat{w}_1\hat{w}_2 = 213$ *will go from* $r = 2$ *to the first symbol of* W, *and* $\check{W} = \check{w}_0\check{w}_1\check{w}_2 = 313$ *will connect the last state in* W *to* $s = 3$. *Combining these we see that*

$$W^* = 13(12213)(12213)\ldots(12213)1$$

and that $2W^*3$ *has all legal transitions joining 2 to 3 regardless of how*

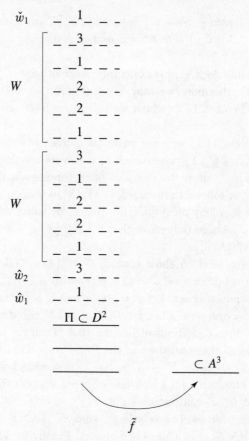

Fig. 16.4. Leaving D^2 and entering A^3 via legal transitions

many repetitions of $W = (12213)$ we include. In Figure 16.4, we illustrate the stacking of levels above D^2 to enter A^3 with legal transitions from this example.

We now complete the proofs of the ergodic genericity results (Theorems 15.1 and 15.2). We combine the approximations of Lemmas 16.2 and 16.3 with the End Preserving Lusin Theorem (Theorem 14.9) and apply the Baire Category Theorem to prove

Theorem 15.1 *If $(E, \mathcal{Q}, \mu^*, \sigma)$ is incompressible then the ergodic homeomorphisms of the sigma compact connected manifold (X, μ) with end action σ and identically zero end charge, i.e., $\mathcal{E} \cap \mathcal{M}_\sigma^0[X, \mu]$, form a*

compact-open topology dense G_δ subset of $\mathcal{M}_\sigma^0[X, \mu]$, the set of homeomorphisms $h \in \mathcal{M}[X, \mu]$ with $h^* = \sigma$ and $c_h \equiv 0$.

Proof We assume $\mathcal{M}_\sigma^0[X, \mu]$ is nonempty, since otherwise the statement is vacuous. Furthermore we may assume $\mu(X) = \infty$, because when $\mu(X) < \infty$, Theorem 15.2 (which we will prove next) gives a stronger result.

Using Theorem 11.1, we may write the subset of ergodic automorphisms $\mathcal{E}$ as $\mathcal{E} = \bigcap_{i=1}^\infty \mathcal{E}_i$, where each $\mathcal{E}_i$ is a weak topology dense open subset of $\mathcal{G}[X, \mu]$. Since the compact-open topology is finer than the weak topology, it follows that each $\mathcal{E}_i \cap \mathcal{M}_\sigma^0[X, \mu]$ is open in the relative compact-open topology on $\mathcal{M}_\sigma^0[X, \mu]$. Hence, by using the Baire Category Theorem, we need only show that $\mathcal{E}_i \cap \mathcal{M}_\sigma^0[X, \mu]$ is a compact-open dense set in $\mathcal{M}_\sigma^0[X, \mu]$

Thus we only need to show that $\mathcal{E}_i \cap \mathcal{M}_\sigma^0[X, \mu] \cap \mathcal{C} \neq \emptyset$ for any compact-open neighborhood $\mathcal{C} = \mathcal{C}(h, K, \epsilon)$ with $h \in \mathcal{M}_\sigma^0[X, \mu]$ and K an h-moving separating set. Let $\delta = \omega(\epsilon)$, where ω is the uniform modulus of continuity of h on K. Let $f \in \mathcal{G}[X, \mu]$ be the ergodic approximant satisfying the three conditions of Lemma 16.3. Then the automorphism $g = h^{-1}f$ satisfies the conditions of the End Preserving Lusin Theorem (Theorem 14.9). Let $h' : X \to X$ be the μ-preserving homeomorphism produced by Theorem 14.9 which has compact support, belongs to the weak open set $h^{-1}\mathcal{E}_i$ and satisfies $d(x, h'(x)) < \delta$ for all x in K. The homeomorphism hh' is obviously in $\mathcal{E}_i$, since $h' \in h^{-1}\mathcal{E}_i$. Also hh' belongs to $\mathcal{C} = \mathcal{C}(h, K, \epsilon)$ by choice of $\delta = \omega(\epsilon)$. Finally by Theorem 14.22, the homeomorphism hh' belongs to $\mathcal{M}_\sigma^0[X, \mu]$ (has zero charge) because h' has compact support. Thus $\mathcal{E}_i \cap \mathcal{M}_\sigma^0[X, \mu] \cap \mathcal{C}$ is not empty. $\qquad \square$

Remark In the special case that $\mu(X) < \infty$, our proof above actually proves more than what is stated above for ergodicity. Namely our use of Lemma 16.2 actually proves:

Theorem 15.2 *Assume $\mu(X) < \infty$. Then for any conjugate invariant property which defines a dense G_δ subset $\mathcal{F}$ of $\mathcal{G}[X, \mu]$ with respect to the weak topology, $\mathcal{F} \cap \mathcal{M}[X, \mu]$ is a dense G_δ subset of $\mathcal{M}[X, \mu]$ with respect to the compact-open topology.*

Proof Since $\mathcal{F}$ and $\mathcal{E}$ are dense G_δ subsets of $\mathcal{G}[X, \mu]$ in the weak topology, so is $\mathcal{F} \cap \mathcal{E}$. Consequently $\mathcal{F}$ contains an ergodic automorphism $\theta \in \mathcal{G}[X, \mu]$ and consequently the whole conjugacy class of θ. Thus in choosing an ergodic approximation f to h in the previous proof we choose f to be the conjugate of $\theta \in \mathcal{F}$ (Lemma 16.2). The remainder of the proof is now the same as above. $\qquad\square$

17

Other Properties Typical in $\mathcal{M}[X, \mu]$

17.1 A General Existence Result

We have seen earlier in Theorem 15.6 that every sigma compact manifold X supports an ergodic μ-preserving homeomorphism, for any OU measure μ. In this chapter, Theorem 17.1, we show that the property of ergodicity may be replaced by any other measure theoretic property $\mathcal{F}$ which is typical in $\mathcal{G}[X, \mu]$ with respect to the weak topology. The results presented here may also be considered as an analog of the results of the second part of Chapter 12 generalizing from R^n to arbitrary sigma compact manifolds. However, unlike the results for R^n, we will not show that $\mathcal{F}$ is generic in the full space $\mathcal{M}[X, \mu]$, but only in the closed subspace $\mathcal{M}_{id}^0[X, \mu]$ consisting of end preserving, zero charge homeomorphisms. (Of course for R^n these are the same as $\mathcal{M}[R^n, \lambda] = \mathcal{M}_{id}^0[R^n, \lambda]$.) The existence result follows, since the subspace $\mathcal{M}_{id}^0[X, \mu]$ contains the identity homeomorphism – and hence is nonempty. We do not know whether the results of this section can be extended to subspaces $\mathcal{M}_\sigma^0[X, \mu]$, for other incompressible end homeomorphisms σ, as was done for the specific property of ergodicity.

One particular generic property $\mathcal{F}$ in $\mathcal{G}[X, \mu]$ is *weak mixing*: we say that $g \in \mathcal{G}[X, \mu]$ is weak mixing on the sigma finite measure space (X, μ) if $g \times g$ is ergodic on $X \times X$ with respect to $\mu \times \mu$. Another such property is the set of zero entropy ergodic automorphisms of an infinite measure space (see Krengel, [79]); Krengel defined the entropy of an infinite measure preserving ergodic automorphism $g \in \mathcal{G}[X, \mu]$ to be the entropy of g_A, the first return map to A $(g_A \in \mathcal{G}[A, \mu_A])$, where A is any finite μ-measured set. See [50], [51], and [1] for other generic properties in $\mathcal{G}[X, \mu]$ when (X, μ) is an infinite measure space. Recall that $\mathcal{M}_{id}^0[X, \mu]$ is the space of all μ-preserving homeomorphisms of X which fix the ends of X and induce an identically zero charge on these ends. This space is nonempty as it always contains the identity homeomorphism. The main result of this chapter is the following theorem proved by the authors and Jal Choksi [18].

Theorem 17.1 *Let μ be an OU measure on a sigma compact, connected manifold X. Let $\mathcal{F}$ be a conjugate invariant subset of $\mathcal{G}[X, \mu]$ which is dense and G_δ with respect to the weak topology. Then $\mathcal{F} \cap \mathcal{M}_{id}^0[X, \mu]$ is a dense G_δ subset of $\mathcal{M}_{id}^0[X, \mu]$ with respect to the compact-open topology. In particular there is a μ-preserving homeomorphism of X which belongs to the set $\mathcal{F}$.*

Note that the denseness of $\mathcal{F}$ in $\mathcal{G}[X, \mu]$ in the weak topology follows if $\mathcal{F}$ contains an antiperiodic automorphism f, because Corollary A1.16 implies that the conjugates of f (in $\mathcal{G}[X, \mu]$) are dense in the weak topology.

17.2 Proof of Theorem 17.1

We devote this section to proving the general existence result cited above. First we note that we need only the consider the case when $\mu(X) = \infty$ because when $\mu(X)$ is finite we have already proved this result in Theorem 15.2. As in the previous chapter the first step involves approximating a homeomorphism $h \in \mathcal{M}_{id}^0[X, \mu]$ by an end preserving ergodic automorphism $f \in \mathcal{F}$, which is uniformly close on a separating compact set K, and setwise the same as h on each component $P(K)$, $P \in \mathcal{P}_K$, of $X - K$. The remaining step involving the Lusin approximation will then be the same as in the last chapter.

To accomplish the first step, we will need the measure theoretic approximation proved in Appendix 1 as Corollary A1.16 which we restate below using different names.

Theorem 17.2 *Let $\tau, g \in \mathcal{G}[X, \mu]$ be any pair of ergodic automorphisms of the infinite sigma finite measure space (X, μ). Let F be any measurable set such that*

$$\mu(X - (F \cup gF)) = \infty.$$

Then there is some conjugate τ' of τ (that is, $\tau' = \gamma^{-1}\tau\gamma$ for some $\gamma \in \mathcal{G}[X, \mu]$) such that $\tau'(x) = g(x)$ for μ-a.e. x in F.

We apply this directly to obtain our main approximation theorem

Theorem 17.3 *Let h be an ergodic end preserving homeomorphism in $\mathcal{M}[X, \mu]$. Let K be a separating compact subset of X and $B(K)$ any connected component of $X - K$ of infinite measure. Let $\tau \in \mathcal{G}[X, \mu]$ be any ergodic automorphism of X.*

Then there is an automorphism $f \in \mathcal{G}[X, \mu]$ conjugate to τ such that $f(x) = h(x)$ for μ-a.e. $x \in X - B(K)$. In particular, $f(x) = h(x)$ for μ-a.e. $x \in K$ and $f(P(K)) = h(P(K))$ for every component $P(K)$, $P \in \mathcal{P}_K$, of $X - K$.

Proof Since h is end preserving and $\mu(B(K)) = \infty$ it follows from Lemma 14.12 that

$$\mu(B(K) \cap hB(K)) = \infty,$$

and consequently for $F = X - B(K) (= \widetilde{B(K)})$, we have

$$\begin{aligned} \mu(X - (F \cup h(F))) &= \mu(X - (\widetilde{B(K)} \cup h(\widetilde{B(K)}))) \\ &= \mu(B \cap h(B)) \\ &= \infty. \end{aligned}$$

Hence we may apply Theorem 17.2 to $g = h$, $F = X - B(K)$ and τ. This yields an automorphism $f = \tau'$ which is conjugate to τ, and satisfies $f(x) = h(x)$ for μ-a.e. x in $X - B(K)$. Since $K \subset X - B(K)$, this proves that f and h agree pointwise (a.e.) on K. For all components $P(K)$ of $X - K$ other than $B(K)$, we have $f = h$ pointwise on $P(K)$, so certainly also setwise. Finally, since f and h are automorphisms with $f(X - B(K)) = h(X - B(K))$, we must also have $f(B(K)) = h(B(K))$, so that they agree setwise on the remaining component $B(K)$. □

We are now in a position to prove our main theorem.

Proof of Theorem 17.1 Write $\mathcal{F} = \bigcap_{i=1}^{\infty} \mathcal{F}_i$ where each $\mathcal{F}_i$ is open and dense in $\mathcal{G}[X, \mu]$ in the weak topology. It remains only to show that $\mathcal{F}_i \cap \mathcal{M}_{id}^0[X, \mu]$ is dense and open in $\mathcal{M}_{id}^0[X, \mu]$ with the compact-open topology. The set is clearly open since the compact-open topology is finer than the weak topology. To show denseness, it is enough to show that $\mathcal{F}_i \cap \mathcal{M}_{id}^0[X, \mu] \cap \mathcal{C}(h, K, \epsilon)$ is nonempty, where $\mathcal{C} = \mathcal{C}(h, K, \epsilon)$ is a compact-open basic neighborhood of some homeomorphism $h \in \mathcal{M}_{id}^0[X, \mu]$, where K is a separating compact set. By Theorem 15.1 we may assume that h is ergodic. Since $\mu(X) = \infty$ one of the components of $X - K$, call it $B(K)$, must have infinite measure.

Observe that both the set $\mathcal{E}$ of ergodic automorphisms (by Theorem 11.1) and the set $\mathcal{F}$ (by hypothesis) are dense and G_δ in $\mathcal{G}[X,\mu]$ with respect to the weak topology. It follows from the Baire Category Theorem that $\mathcal{F} \cap \mathcal{E}$ is also dense and G_δ. In particular, $\mathcal{F} \cap \mathcal{E}$ is not empty, so we pick some automorphism $\tau \in \mathcal{E} \cap \mathcal{F}$. Apply the above Theorem 17.3 to h, $\tau \in \mathcal{E} \cap \mathcal{F}$, K and $B(K)$ as above to obtain f. Since $\mathcal{F}$ is conjugate invariant and f is a conjugate of $\tau \in \mathcal{F}$, it follows that f belongs to $\mathcal{F}$ and consequently each $\mathcal{F}_i$. The End Preserving Lusin Theorem (Theorem 14.9) is now applied to $g = h^{-1}f \in h^{-1}\mathcal{F}_i$, a weak topology open set, with $\delta = \omega(\epsilon)$ where ω is the modulus of continuity of h on K. This produces a compactly supported μ-preserving homeomorphism h' which also belongs to $h^{-1}\mathcal{F}_i$, and satisfies $h'(K) = K$ and $d(h'(x), x) < \delta$ for all x in K. Then the verification that $hh' \in \mathcal{F}_i \cap \mathcal{M}^0_{h*}[X,\mu] \cap \mathcal{C}(h, K, \epsilon)$ is the same as we have seen in the proof of Theorem 15.2. $\quad\square$

It is worth noting that the above proof requires only that there is at least one isolated end of infinite measure which is fixed by h^*.

17.3 Weak Mixing End Homeomorphisms

We have shown above that *weak mixing is typical in* $\mathcal{M}^0_{id}[X,\mu]$, *or more generally (see the comment after the last proof) in* $\mathcal{M}^0_\sigma[X,\mu]$ *when* σ *fixes an end of infinite measure.* In this section we show the same result for $\mathcal{M}_\sigma[X,\mu]$ when σ is componentwise weak mixing. We recall from Chapter 14 that this means that for any clopen sets U, V in E^∞ (the ends of infinite measure, with the relative topology) the set $\{k : \sigma^k U \cap V \neq \emptyset\}$ contains arbitrarily long strings of consecutive integers (integer intervals). The following theorem was proved by the authors in [21].

Theorem 17.4 *Let* μ *be an OU measure on a sigma compact, connected manifold* X. *Let* $\mathcal{F}$ *be a conjugate invariant subset of* $\mathcal{G}[X,\mu]$ *which is dense and* G_δ *with respect to the weak topology. Assume that* σ *is componentwise weak mixing on* E^∞. *Then* $\mathcal{F} \cap \mathcal{M}_\sigma[X,\mu]$ *is dense and* G_δ *in* $\mathcal{M}_\sigma[X,\mu]$ *with respect to the compact-open topology.*

Proof The overall structure of the proof is the same as that of Theorem 17.1, so we indicate only the differences. Let $\mathcal{C}(h, K, \epsilon)$ be the compact-open neighborhood of that proof, with $h \in \mathcal{M}_\sigma[X,\mu]$ and K a compact separating set. Since σ is componentwise weak mixing on E^∞, it is incompressible and has zero charge on E. Hence we can assume, just as in the proof of Theorem 17.1, that h is ergodic. Instead of using Theorem 17.2 to approximate h, we use Theorem A1.17, with $X_1, \ldots, X_m$ as the components of $X - K$ of infinite μ-measure. In order to assure that the 0–1 matrix $\mathbf{T} = \mathbf{T}(h, K)$, defined by $t_{ij} = 1$ if $\mu(hX_i \cap X_j) = \infty$, is mixing, we use the assumption that h^* is componentwise weak mixing on E^∞. According to Theorem 14.19, this is enough to guarantee that the matrix $\mathbf{T}$ is mixing, so that Theorem A1.17 can be applied. $\quad\square$

Of course the above result does not apply when X has finitely many ends, since in that case the end homeomorphism is a permutation, and cannot be topologically weak mixing. So we don't know, for example, if there is a weak mixing volume preserving homeomorphism of the infinite strip $R^1 \times [0, 1]$ which transposes its two ends. See however [31] for a related result.

We now give an example of a system (X, μ, h) where h^* is componentwise weak mixing on the ends of X (all the ends have infinite measure). As in Example 14.5, define the sigma compact manifold X as $D - C$, where D is the closed unit disk $\{(x, y) : x^2 + y^2 \leq 1\}$ and C is the standard Cantor ternary set lying on the line $I = [-1/2, 1/2]$ along the x-axis. The set C may be naturally identified with the ends E of X, noting that the end topology on E is the same as the relative topology of C as a subset of the disk D. Let μ be any of the OU measures on X for which all the ends have infinite measure. Let σ be any homeomorphism of the set $C = E$ which is componentwise weak mixing. For example, the two sided shift when C is viewed as the countable product of a two symbol set. Antoine [33] proved that any homeomorphism of C can be extended to a homeomorphism of D (see also the extension theorems of Keldyš [77] or Oxtoby [90] for dimensions $n > 2$). The result of Berlanga and Epstein, Theorem A2.8, ensures that we can assume that the extension h of σ preserves μ and satisfies $h^* = \sigma$.

17.4 Maximal Chaos on Noncompact Manifolds

In this section we apply Theorem 17.1 to demonstrate that any sigma compact manifold supports a maximally chaotic homeomorphism. This section is an extension of the results of Section 4.5 to the noncompact case. The following result is new, though similar to Theorem 7 of [29]. The technique is similar to that used in [2] with transitivity replaced by weak mixing. This is the explicit version of the result informally referred to as Theorem I in Section 11.3.

Theorem 17.5 *Maximally chaotic homeomorphisms are dense in the nonempty set $\mathcal{M}_{id}^0[X, \mu]$, with respect to the compact-open topology. In particular every sigma compact connected manifold X supports a maximally chaotic homeomorphism which moreover preserves a given OU measure.*

Proof Let (U_k, V_k, W_k), $k = 1, \ldots, \infty$, be an enumeration of all triples of a countable basis of open subsets of X. For simplicity, we may assume that these open sets have compact closures. Let $h \in \mathcal{M}_{id}^0[X, \mu]$ be given, together with a compact set $K \subset X$ and a positive number δ. We will construct a homeomorphism $f \in \mathcal{M}_{id}^0[X, \mu]$ with $\|f\|_K < \delta$, such that for every k there are fh-periodic points p_k and q_k in U_k and a positive integer m_k with $(fh)^{m_k}(p_k) \in V_k$ and $(fh)^{m_k}(q_k) \in W_k$. This condition clearly implies that the periodic points of fh are dense, that fh is transitive (in fact it is 'topologically weak mixing'), and following the same reasoning as in the proof of Theorem 4.8 for the compact case (see also Theorem 4.5), that fh has maximal sensitivity on initial conditions (condition (iii) of maximal chaos in Definition 4.7).

By the Theorem 17.1, there is a $g_1 \in \mathcal{M}_{id}^0[X, \mu]$ with $U(g_1, identity)$ arbitrarily small, such that $g_1 h$ is weak mixing (note $U(f, g)$ is the metric on $\mathcal{M}$ giving the compact-open topology in Chapter 11). It follows that there exist distinct points p_1 and q_1 in U_k such that $(g_1 h)^{m_1}(p_1) \in V_1$ and $(g_1 h)^{m_1}(q_1) \in V_1$ for some m_1, and $(g_1 h)^{u_1}(p_1)$ and $(g_1 h)^{v_1}(q_1)$ are arbitrarily close to p_1 and q_1, for positive integers u_1 and v_1, respectively. Hence by Theorem 2.4 we can find an arbitrarily small (close to identity) homeomorphism f_1 which maps $(g_1 h)^{u_1}(p_1)$ to p_1 and $(g_1 h)^{v_1}(q_1)$ to q_1, and hence for which p_1 and q_1 are periodic with respect to $f_1 g_1 h$. Furthermore we can perform this construction so that any given finite set is fixed by $f_1 g_1$ by requiring that f_1 take $g_1(a)$ into a for any point a in this given finite set. We continue recursively as in the proof of Theorem 4.8, at each step fixing the periodic orbits of the points $p_1, q_1, p_2, q_2, \ldots, p_{i-1}, q_{i-1}$. $\qquad\square$

Appendix 1
Multiple Rokhlin Towers and Conjugacy Approximation

A1.1 Introduction

The results presented in this chapter are purely measure theoretic and do not rely on any manifold structure on the underlying space. This is the reason we have grouped this material separately as an appendix. Most of this chapter is devoted to presenting the background, proof, and consequences of a result of the first author [13] that we call the Multiple Tower Rokhlin Theorem. The classical work of Rokhlin, Kakutani and Halmos showed that an antiperiodic automorphism of a finite Lebesgue space could be *nearly* represented as a single stack, or 'tower', consisting of disjoint iterates of a 'base' set. The modification presented here as Theorem A1.4 shows that it can be *exactly* represented as a collection of disjoint towers of prescribed heights and measures, as long as the set of heights is relatively prime. This result was in fact motivated by questions related to manifold homeomorphisms, and indeed corollaries of this tower theorem are applied several times in the main text. An infinite measure analog of our tower theorem established by the authors and Jal Choksi, our Theorem A1.15 and Corollary A1.16, also has important implications for manifold homeomorphisms preserving an infinite measure. That application forms the basis of Chapter 17.

This chapter is organized as follows. Section A1.2 reviews the basic 'skyscraper' construction developed by Kakutani [75] and others. Elements of this construction and the related 'induced transformation' are applied directly (that is, without the further conjugacy approximation results) in Chapters 7 and 16 to produce, respectively, ergodic homeomorphisms on the cube and on any sigma compact manifold. Section A1.3 presents the motivation and proof for the Multiple Tower Rokhlin Theorem, our Theorem A1.4, which we described above. Sections A1.4–

A1.6 apply Theorem A1.4 to obtain various 'conjugacy approximation' theorems. These results state, in various forms, that a given automorphism h in $\mathcal{G}[X,\mu]$ can be approximated by some conjugate $\hat{g}$ of a given antiperiodic automorphism g. This means that $\hat{g}$ is of the form $f^{-1}gf$, for some automorphism f in $\mathcal{G}[X,\mu]$. The classical result of this type is due to Halmos, who showed that h could be approximated by $\hat{g}$ in the weak topology. This means that $\hat{g}$ is close to h setwise on a finite family of sets. In section A1.4 we apply the Multiple Tower Rokhlin Theorem to find conditions (Theorem A1.8) under which $\hat{g}$ can be made to equal h pointwise on a given set F. We apply this to show (Theorem 8.4) that the conjugates of an antiperiodic automorphism are dense in $\mathcal{G}[I^n,\mu]$ in the uniform topology (strictly stronger than the weak topology of Halmos's result). In section A1.5 we apply the previous results to show (Theorem A1.9) that if g is antiperiodic we can find a partition of the underlying space such that the transition probabilities with respect to g equal any given set p_{ij} of probabilities which form a mixing stochastic matrix. In section A1.6 we prove several conjugacy results which find conditions under which $\hat{g}$ can be made to exactly agree with h setwise on a given finite algebra of sets, while also being uniformly close to h (with respect to some metric) on a certain set. All the results described so far are concerned with finite Lebesgue spaces. The final section of this chapter, A1.7, gives some related constructions for infinite Lebesgue spaces. It gives a sufficient condition (Corollary A1.16) for some conjugate $\hat{g}$ to agree pointwise with a given ergodic automorphism h on a given set F. This condition is weaker than an earlier one established by Choksi and Kakutani [50].

A1.2 Skyscraper Constructions

The results of this section do not require that $\mu(X) < \infty$; that is they apply to both finite or sigma finite Lebesgue spaces. There are two kinds of skyscraper constructions: the first starts with a μ-recurrent automorphism g of (X,μ), and a measurable set $A \subset X$, and then represents g as a skyscraper over A with induced automorphism g_A; the second starts with a 'primitive automorphism' g_A of A and builds automorphisms g of X (over A) having induced automorphism g_A – these g's are called skyscrapers (or towers) over the primitive g_A. These are standard constructions in ergodic theory and can be found in the books [63], [93], and [102]. Section 7.1 uses and motivates the general constructions of this section.

Given a recurrent automorphism $g \in \mathcal{G}[X, \mu]$, and a *sweep-out* set A (i.e., $\mu\left(X - \bigcup_{i=0}^{\infty} g^i A\right) = 0$), one may generate a *skyscraper over A* as follows: By the assumption of recurrence, almost every point $x \in A$ has a finite *return time* $k \geq 1$, defined as the least positive iterate of g which returns x to A. Define A^k, $k = 1, \ldots, \infty$, to be the subset of A with return time k. Since A is a sweep-out set, almost every point in X lies in some set $g^{i-1}(A)$, with minimal positive i, and hence belongs to some unique set $g^{i-1}\left(A^k\right)$. Call the set of such points $A_{k,i}$ and note that in this notation we have $A^k = A_{k,1}$. For each $i = 1, 2, \ldots$, the set $F_i = \bigcup_{k=1}^{\infty} A_{k,i} (= \bigcup_{k=1}^{\infty} g^{i-1} A^k)$ is called *the ith floor of the skyscraper* (the exponent $i - 1$ ensures that base set A has index $i = 1$ and can be called 'the first floor' as in the American numbering of floors of buildings). The triangular partition $\mathcal{A} = \{A_{k,i} : k = 1, \ldots, \infty; i = 1, \ldots, k\}$ is called *the skyscraper (partition) for g based on A*. The set $A = \bigcup_k A^k$ is called the *base* of the skyscraper and the set $g^{-1}(A) = \bigcup_k A_{k,k}$ is called the *top* of the skyscraper. The set $A_k = \bigcup_{i=1}^{k} g^{i-1}\left(A^k\right) = \bigcup_{i=1}^{k} A_{k,i}$ is called the *column of height k*. The distribution of the measures of the columns of a skyscraper $\mathcal{A} = \{A_{k,i}\}$ is denoted by $d(\mathcal{A}) = (\mu(A_1), \mu(A_2), \ldots)$.

From this skyscraper partition of X we can obtain a geometric picture of the action of g on (X, μ). Consider the *first return map* to A, denoted g_A, which is the μ-preserving automorphism of (A, μ) defined by

$$g_A(x) = g^k(x), \; x \in A^k$$

where k is the smallest positive integer such that $g^k(x) \in A$. The automorphism g_A is also referred to as the *induced automorphism on A*, or the *primitive map on A*. Note that the (possibly) denumerably many columns A_k of height k form a partition of the space X. Any point x in the kth column A_k, which is not in the top of the skyscraper (i.e., $x \in A_{k,i}$, $i < k$) maps directly to the 'point above it' in $A_{k,i+1}$ (i.e., to $g(x) \in A_{k,i+1}$). If $x \in A_{k,k}$ is a point in the top of the skyscraper then $y = g^{-(k-1)}(x)$ is the point in the base directly below x, and it is easy to see that g maps x to the point in the base $g_A(y)$. If we denote by μ_A the restriction of μ to subsets of A, then g_A is a measure preserving automorphism of (A, μ_A) which is ergodic whenever g is:

Theorem A1.1 *Let (X, μ) be a finite or sigma finite Lebesgue space. Given a μ-recurrent automorphism $g \in \mathcal{G}[X, \mu]$ and a set $A \subset X$ of positive μ-measure, the induced automorphism g_A on (A, μ_A) is a μ_A-preserving automorphism of A. Furthermore, g_A is ergodic on (A, μ_A)*

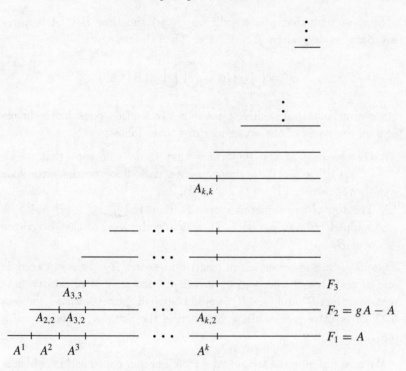

Fig. A1.1. Skyscraper partition $\mathcal{A} = \{A_{k,i}\}$ with columns A_k over $A^k \subset A$

when g is ergodic on (X, μ). (Here μ_A denotes the measure μ restricted to the measurable subsets of A.)

Proof Note that for any measurable set $B \subset A = \bigcup_k A^k$,

$$
\begin{aligned}
\mu_A(g_A(B)) &= \mu_A\left(\bigcup_{k=1}^{\infty} g^k(B \cap A^k)\right) \\
&= \sum_{k=1}^{\infty} \mu_A(g^k(B \cap A^k)) \\
&= \sum_{k=1}^{\infty} \mu_A(B \cap A^k) \\
&= \mu_A(B),
\end{aligned}
$$

which shows that g_A is a measure preserving automorphism of (A, μ_A).

Suppose now that g is ergodic on (X, μ) and that $B \subset A$ is a g_A-invariant set of positive μ_A-measure. Then the set

$$B^* = \bigcup_{i=0}^{\infty} g^i(B) = \bigcup_{k=1}^{\infty} \bigcup_{i=0}^{k-1} g^i(B \cap A^k)$$

is a g-invariant set of positive μ-measure. To see this, consider the image of g on the floors of the skyscraper over B as follows:

(i) On the part of the skyscraper that is not the top (that is, on $\bigcup_{k=2}^{\infty} \bigcup_{i=0}^{k-2} g^i(B^k)$), g simply moves each floor to the next floor above (namely $\bigcup_{k=2}^{\infty} \bigcup_{i=1}^{k-1} g^i(B^k)$).

(ii) The top of the skyscraper over B, namely $\bigcup_{k=1}^{\infty} g^{k-1}(B \cap A^k)$, is mapped by g to $g_A(B) = B$, which is the base of the skyscraper over B.

Thus B^* is a g-invariant set of positive measure. If g_A was not ergodic and B and $A - B$ were two g_A-invariant subsets of A of positive μ_A-measure, then B^* and $(A-B)^*$ would be two disjoint g-invariant subsets of X of positive μ-measure, contradicting the assumed ergodicity of g on (X, μ). $\qquad\square$

We now describe the second type of skyscraper construction, which is in some sense a reverse of the previous construction.

Suppose that we have a measure space (A, μ) and a μ-recurrent automorphism $g \in \mathcal{G}[A, \mu]$. Let $\{A_i : i = 0, 1, \ldots, \infty\}$ be a decreasing sequence of subsets of A where $A = A_0 \supset A_1 \supset A_2 \supset \cdots$ and $\mu(A_i)$ decreases to 0. Consider the measure space $\tilde{X}$ which is the disjoint union of A_i's; i.e., for each i consider the copy of A_i given by $\tilde{A}_i = (A_i, i)$, so that

$$\tilde{X} = \bigcup_{i=0}^{\infty} (A_i, i).$$

Identifying A_i with $\tilde{A}_i$, $\tilde{X}$ inherits a sigma algebra and measure $\tilde{\mu}$ from (A, μ) in the obvious way. Note that $\tilde{\mu}$ is finite or infinite as $\sum_i \mu(A_i)$ converges or diverges. Define the automorphism $\tilde{g}$ of $(\tilde{X}, \mu)$ by

$$\tilde{g}(x, i) = \begin{cases} (x, i+1) & \text{when } x \in A_{i+1} \\ (g(x), 0) & \text{when } x \notin A_{i+1}. \end{cases}$$

Abusing notation, and identifying A_0 as a subset of $\tilde{X}$, we note that the first return map $\tilde{g}_{A_0}$ is just g and in this sense this construction reverses the process of inducing on a subset. Note further that the ith floor of

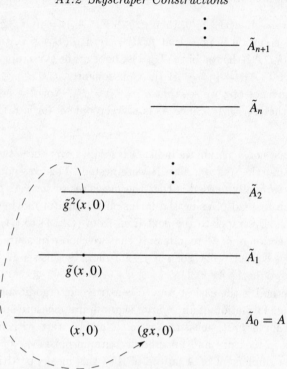

Fig. A1.2. $\tilde{g}$, the skyscraper over g on A

the skyscraper is just the set A_{i-1} (or more precisely its copy $\tilde{A}_{i-1}$). We call $\tilde{g}$ the skyscraper over g on A.

The following result has already been used in Section 7.1, where a special case was proved.

Theorem A1.2 *If g is an ergodic measure preserving automorphism of (A, μ) then the skyscraper $\tilde{g}$ over g on A is ergodic.*

Proof The fact that $\tilde{g}$ is measure preserving is easy and left to the reader. So suppose that g is ergodic. Since $\mu(A_i)$ decreases to 0, it follows easily that the $\tilde{g}$-orbit of almost every point in $\tilde{A}_i$ leaves $\tilde{A}_i$ and enters A_0. Consequently it is easy to see that

$$\tilde{X} = \bigcup_{i=1}^{\infty} \tilde{g}^{-i} A_0$$

(i.e., A_0 is a sweep-out set for $\tilde{g}$). Let B be a $\tilde{g}$-invariant set. Set

$B_0 = B \cap A_0$. If $\mu(B_0) = 0$, then $\mu(\tilde{g}^{-i}(B \cap A_0)) = \mu(B \cap \tilde{g}^{-i}A_0)$ and the above equation implies that $\mu(B) = 0$. If $\mu(B_0) > 0$, then setting $C_i = \tilde{g}^{-1}B_0 \cap \tilde{A}_i$ (the top of the $(i+1)$st floor of the skyscraper), we have that $g^{-1}(B \cap A_0) = \bigcup_0^\infty \tilde{g}^{-i}(C_i)$. Consequently $g^{-1}(B_0) \subset B_0$; using the μ-recurrence of g we see that $g^{-1}(B_0) = B_0$. Now the ergodicity of g implies that $B_0 = A_0$. Since A_0 is a sweep-out set for $\tilde{g}$, it follows that $B = \tilde{X}$. $\square$

In the last construction we built a skyscraper over the set A_0, and we assumed that the sets $A_0, A_1, \ldots$, were nested. This was not necessary – all that we really needed is that the measures of the sets $\mu(A_i)$ tend to 0. Then instead of using the inclusion relation in moving from one floor of the skyscraper to the next (i.e., from $((A_j, j)$ to $(A_{j+1}, j+1))$ under the action of $\tilde{g}$, we require only that we choose measure preserving embeddings $\phi_{ij} : A_i \to A_j$ for $i > j$ – for more details see K. Petersen's book on ergodic theory [93].

This method is the easiest way to construct an ergodic measure preserving transformation of an *infinite* sigma finite measure space (X, μ), given one for a *finite* measure space. Namely start with any subset A of finite μ-measure and an ergodic automorphism g_A of A. Then 'stack the complement of A on top of A' – this means partition $X - A$ into $A_1, A_2, \ldots$ where the sequence $\mu(A_n)$ goes to zero and is bounded above by $\mu(A)$ – then letting $A_0 = A$, apply the above construction to the sets A_i. Because the induced automorphism g_A is ergodic on A, any skyscraper over A will be an ergodic automorphism of the space (X, μ). This construction is used in Chapters 16 and 17 and mentioned in Chapter 12.

A1.3 Multiple Tower Rokhlin Theorem

For probability spaces (X, μ), the classical Rokhlin Tower Theorem [100] says (see also Kakutani's paper [75], and Halmos's book [72]):

Theorem A1.3 *Given any antiperiodic automorphism $g \in \mathcal{G}[X, \mu]$, a positive integer m, and $\epsilon > 0$, there is a measurable set B such that the sets $B, g(B), \ldots, g^{m-1}(B)$ are disjoint and have combined measure at least $1 - \epsilon$.*

This is clearly equivalent to requiring that the combined measure of $B, g(B), \ldots, g^{m-1}(B)$ is *exactly* $1 - \epsilon$, by selecting an appropriately measured subset of B. The latter formulation may be expressed in terms of skyscrapers by saying that the space X can be partitioned into two columns for g, one of height m occupying μ-measure $1 - \epsilon$ and the other column of height 1 and measure ϵ; i.e., there is a skyscraper $\mathcal{A} = \{A_{k,i}\}$ for g with column distribution $d(\mathcal{A}) = (\epsilon, 0, 0, \ldots, 1 - \epsilon, 0, 0, \ldots)$, that is, with $\mu(A_1) = \epsilon$ and $\mu(A_m) = 1 - \epsilon$. The main result of this chapter, the Multiple Tower Rokhlin Theorem, is a generalization of Rokhlin's Tower Theorem that guarantees a skyscraper with any column distribution for which *the heights of the columns of positive measure are relatively prime* (columns 1 and m in Rokhlin's Theorem). This result was first proved for *finite* distributions, which is sufficient for the applications in Part I involving homeomorphisms of compact manifolds ([11]). The version stated below and proved in this chapter is needed for applications to homeomorphisms of noncompact manifolds. This theorem was first proved by S. Alpern and the proof we present here is from [13] (see also [23], [18], [57], [98] for related work). A further generalization to nonsingular (rather than measure preserving) transformations can be found in [24]. The main result of this section is the following:

Theorem A1.4 (Multiple Tower Rokhlin Theorem) *Let* $g \in \mathcal{G}[X, \mu]$ *be any antiperiodic automorphism of a nonatomic Lebesgue probability space* (X, Σ, μ). *Then given any countable probability distribution* $\pi = (\pi_1, \pi_2, \ldots)$, *for which the k's with $\pi_k > 0$ are relatively prime, there is a skyscraper partition* $\mathcal{P} = \{P_{k,i}\}$, $k = 1, \ldots, \infty$, $i = i, \ldots, k$, *of X satisfying*

(i) $P_{k,i} = g^{i-1}(P_{k,1})$

(ii) $\mu \left(\bigcup_{i=1}^{k} P_{k,i} \right) = \pi_k$.

The proof of the Multiple Tower Rokhlin Theorem involves some special notation, for which we fix the antiperiodic automorphism g and the skyscraper column distribution π given in the theorem. Let S denote the (relatively prime) set of k's with $\pi_k > 0$. The constructive techniques of the proof can be applied only to produce skyscrapers with finitely many columns of approximately desired distributions. The required skyscraper will then be obtained as a limit of the finite skyscrapers, with respect to a complete metric on skyscraper partitions. The target distributions for the finite skyscrapers will be finite versions of the distribution π, appropriately normalized to sum to 1 ($= \mu(X)$). To produce these finite target distributions, we define

$$S = \{k : \pi_k > 0\},$$

$$S^j = \{k \le j : k \in S\},$$

$$s_j = \sum_{k \in S^j} \pi_k, \text{ and}$$

$$\pi^j = \left(\pi_1^j, \pi_2^j, \ldots, \pi_j^j, 0, 0, \ldots\right), \text{ where } \pi_k^j = \pi_k/s_j, \ k \le j.$$

Since the (potentially infinite) set S consists of integers with greatest common divisor 1, it follows that some finite subset S^{j_0} is also a relatively prime set of numbers. Consequently all integers greater than or equal to some fixed integer L may be represented as positive integer linear combinations of S^{j_0}. The skyscraper partitions (henceforth called simply skyscrapers) approximating the required skyscraper $\mathcal{P}$ will not have the right distribution (π), but they will never have positive measured columns of height k for which $\pi_k = 0$. We will call such a skyscraper a π-*skyscraper*. That is, a π-skyscraper is a measurable partition $\mathcal{R} = \{R_{k,i}\}, k = 1, \ldots, \infty, \ i = 1, \ldots, k$, for which

$$R_{k,i} = g^{i-1}(R_{k,1}), \ i \le k, \text{ and } \pi_k = 0 \Rightarrow \mu(R_{k,1}) = 0.$$

We shall further call $\mathcal{R}$ a π-*skyscraper of type* j, if it has no column heights larger than j (i.e., $\mu(R_{k,1}) = 0$ for $k > j$). The *column of height* k is denoted by

$$R_k = \bigcup_{i=1}^{k} R_{k,i}$$

and the distribution of the measures of the columns is denoted by

$$d(\mathcal{R}) = (\mu(R_1), \mu(R_2), \ldots).$$

Note that in this notation the Multiple Tower Rokhlin Theorem asserts the existence of a π-skyscraper $\mathcal{P}$ with $d(\mathcal{P}) = \pi$.

For any skyscraper $\mathcal{P}$, if $x \in P_{k,i}$ we call (k,i) the $\mathcal{P}$ *label of* x. The proof of the Multiple Tower Rokhlin Theorem will involve the recursive construction of π-skyscrapers $\mathcal{P}^j$ of type j, which will converge to the required π-skyscraper $\mathcal{P}$ with respect to the partition metric

$$\|\mathcal{R} - \mathcal{Q}\| = \mu\{x : x \text{ has different } \mathcal{R} \text{ and } \mathcal{Q} \text{ labels}\}$$

on the complete space of π-skyscrapers. To ensure that

$$d(\mathcal{P}) = \lim_j d(\mathcal{P}^j) = \pi,$$

we use the sum metric on l^∞ for the distributions and observe that

$$|d(\mathcal{R}) - d(\mathcal{Q})| \equiv \sum_{k=1}^{\infty} |\mu(R_k) - \mu(Q_k)| \le 2\|\mathcal{R} - \mathcal{Q}\|.$$

In our construction, we would like to have $d(\mathcal{P}^j) = \pi^j$, but in fact we are only able to ensure that $|d(\mathcal{P}^j) - \pi^j|$ is small, or equivalently that $\triangle_j(\mathcal{P}^j)$ is small, where

$$\triangle_j(\mathcal{R}) = \max_{k \in S^j} \left(1 - \pi_k^j / \mu(R_k)\right).$$

The construction of the skyscrapers $\mathcal{P}^j$ will use the following two lemmas about an antiperiodic automorphism $g \in \mathcal{G}[X, \mu]$.

Lemma A1.5 *For any positive integer m there is a sweep-out set B, all of whose return times are at least m. That is, there is a measurable set B satisfying*

(i) *The sets $B, g(B), \ldots, g^{m-1}(B)$ are disjoint, and*
(ii) *$\bigcup_{i=1}^{\infty} g^i(B)$ is the whole space X.*

Proof Let B be a maximal set satisfying condition (i). If B does not satisfy (ii) then we may adjoin to B the maximal set satisfying property (i) with respect to the complement of the invariant set $\bigcup_{i=1}^{\infty} g^i(B)$. (For more details, see [72, pp. 70–72].)   □

Lemma A1.6 *Let $j \ge j_0$ and $\epsilon > 0$ be given. Then to every π-skyscraper $\mathcal{R}$ of type j, there is another π-skyscraper $\mathcal{Q}$ of type j with*

$$\|\mathcal{Q} - \mathcal{R}\| \le \triangle_j(\mathcal{R}), \quad and \quad |d(\mathcal{Q}) - \pi^j| < \epsilon.$$

Proof Let B be the sweep-out set given by the previous lemma with respect to some large integer m to be specified later. Partition B into sets B^l, $l = 1, 2, \ldots$, so that if $x, y \in B^l$, then x and y have the same return time m_l to B and $g^r(x)$ and $g^r(y)$ belong to the same element of $\mathcal{R}$, for $r = 1, \ldots, m_l - 1$. Next partition each set B^l into sets B_0^l and B_k^l, $k \in S^j$, so that

$$\mu(B_0^l) = \alpha \mu(B^l), \quad and \quad \mu(B_k^l) = \beta_k \mu(B^l),$$

where $\alpha = 1 - \triangle_j(\mathcal{R}) = \min_{k \in S^j}\left(\pi_k^j / \mu(R_k)\right)$, $\beta_k = \pi_k^j - \alpha \mu(R_k)$, and $\alpha + \sum_{k \in S^j} \beta_k = 1$. Note that if the given skyscraper $\mathcal{R}$ already has the distribution π^j then $\triangle_j(\mathcal{R})$ is 0, α is 1, and hence all the β_k are 0. Let C_k^l denote the column based on B_k^l, that is,

$$C_k^l = \bigcup_{r=0}^{m_l-1} g^r\left(B_k^l\right).$$

Then define $D_k = \bigcup_l C_k^l$. Observe that

$$\mu(D_0) = \alpha = 1 - \triangle_j(\mathcal{R}), \ \mu(D_k) = \beta_k, \text{ and} \tag{A1.1}$$

$$\mu(R_k \cap D_0) = \alpha\,\mu(R_k) = \pi_k^j - \beta_k. \tag{A1.2}$$

We now define $\mathcal{Q}$ on D_0 to be the same as $\mathcal{R}$. That is, points in D_0 have the same $\mathcal{Q}$-labels and $\mathcal{R}$-labels. So regardless of how we subsequently define $\mathcal{Q}$ on the complement of D_0, we will have

$$\|\mathcal{Q} - \mathcal{R}\| \le \mu(X - D_0) = 1 - \alpha = \triangle_j(\mathcal{R}),$$

which is the first requirement of the lemma. So we can change as many of the labels (indices of $\mathcal{R}$) on the complement of D_0 (i.e., on $\bigcup_{k \in S^j} D_k$) as needed to ensure the second requirement of the lemma, that the distribution of $\mathcal{Q}$ is near π^j.

If we could define $\mathcal{Q}$ on $X - D_0$ so that $\mu(Q_k \cap D_k) = \mu(D_k)$, we would have by (A1.2) and (A1.1) that

$$\mu(Q_k) = \mu(R_k \cap D_0) + \mu(Q_k \cap D_k) = \pi_k^j, \ k \in S^j$$

and consequently that $d(\mathcal{Q})$ would be exactly π^j. While this is not possible, we can define $\mathcal{Q}$ on the D_k so that $\mu(Q_k \cap D_k)$ is nearly equal to $\mu(D_k)$ and consequently $d(\mathcal{Q})$ is close to π^j, as required by the lemma.

We now define $\mathcal{Q}$ on D_k, for an arbitrary but fixed value k, by specifying it on each column C_k^l as follows. For simplicity of notation we suppress the indices k and l and denote $B_k^l = E$, $m_l = M$, and $C = C_k^l = \bigcup_{r=0}^{M-1} g^r(E)$. Suppose the base of C has $\mathcal{R}$-label (k_1, i_1) and the top (Mth) level has $\mathcal{R}$-label (k_2, i_2). (Formally, this means $E \subset R_{k_1,i_1}$ and $g^{M-1}(E) \subset R_{k_2,i_2}$.) We assign a unique $\mathcal{Q}$-label to each level of the column C, that is, to each of the sets $E, gE, \ldots, g^{M-1}E$. Thus we need to assign M labels to the M levels. Of course we want most of these labels to be of the form $(k, \cdot)$, so that we achieve the result that $\mu(Q_k \cap C)$ is nearly equal to $\mu(C)$, as mentioned above. We will assign the labels to the levels $1, \ldots, M$ in four distinct blocks. Starting from the bottom (level 1), we will call these blocks the Bottom, Middle, Transition, and Top. The Middle block will be the largest and will be labeled with $(k, \cdot)$ to achieve the desired result. A particular instance of our labeling procedure is described after the proof in an example. The reader may wish to look at the example while following the general proof.

We begin by describing the labeling of the Bottom and Top (which is identical to the labeling of $\mathcal{R}$). To ensure that this labeling makes $\mathcal{Q}$ a π-skyscraper, we are forced to retain the $\mathcal{R}$-labels on the lower levels, going up until the label (k_1, k_1) is reached; similarly we must go down from the top until the label $(k_2, 1)$ is reached. Note that in a π-skyscraper the label (k_1, k_1) can be followed by any label with second coordinate 1, and similarly $(k_2, 1)$ can be preceded by any label with equal coordinates. So our problem is to fill in the unlabeled terms $(-)$ in the following list:

$$(k_1, i_1), (k_1, i_1 + 1), \ldots, (k_1, k_1), -, -, \ldots, -, (k_2, 1), \ldots, (k_2, i_2).$$

Now we describe the labeling of the Middle block. We begin by following the label (k_1, k_1) with successive blocks of the form

$$(k, 1), (k, 2), \ldots, (k, k),$$

until there are T unlabeled places remaining between the final (k, k) and the label $(k_2, 1)$, where T satisfies $L \leq T \leq L + k$ and L is the integer defined shortly after the definition of π^j (see the paragraph following the statement of Theorem A1.4). (Recall that the letter k is no longer a variable but a fixed label.) We now describe the labeling of these T levels, which we call the Transition block. The definition of L ensures that T can be written as a positive integer (linear) combination of the elements of S^{j_0}, so that the T unlabeled places can be filled in with blocks of the form

$$(k', 1), (k', 2), \ldots, (k', k'),$$

for varying values of $k' \in S^{j_0}$. This procedure ensures that $\mathcal{Q}$ is a π-skyscraper. Furthermore, of the M labels given to the column C, all but at most

$$(k_1 - i_1 + 1) + T + k_2 \leq j + (L + j) + j = L + 3j$$

have first coordinate k, and correspond to points in Q_k. Consequently

$$\frac{\mu(Q_k \cap C)}{\mu(C)} \geq 1 - \frac{L + 3j}{M} \geq 1 - \frac{L + 3j}{m},$$

where we recall that m comes from Lemma A1.5 and can be as large as we wish – we now specify the size of m. Since the above displayed inequality is true for every column $C = C_k^l$ which constitutes the set $D_k = \bigcup_l C_k^l$, it follows that

$$\frac{\mu\left(Q_k \cap D_k\right)}{\mu\left(D_k\right)} \geq 1 - \frac{L + 3j}{m} \geq 1 - \epsilon/2,$$

if we take $m > 2\left(L + 3j\right)/\epsilon$. Finally, we calculate that for $k \in S^j$,

$$\begin{aligned}
\pi_k^j - \mu\left(Q_k\right) &\leq \pi_k^j - \mu\left(Q_k \cap D_0\right) - \mu\left(Q_k \cap D_k\right) \\
&\leq \pi_k^j - \mu\left(R_k \cap D_0\right) - \left(1 - \epsilon/2\right)\mu\left(D_k\right) \\
&\leq \pi_k^j - \left(\pi_k^j - \beta_k\right) - \left(1 - \epsilon/2\right)\beta_k \\
&\leq \epsilon\beta_k/2,
\end{aligned}$$

where the third inequality uses equation (A1.1) that $\mu(D_k) = \beta_k$. It follows that

$$\begin{aligned}
\left|\pi^j - d\left(\mathcal{Q}\right)\right| &= \sum_{k \in S^j} \left|\pi_k^j - \mu\left(Q_k\right)\right| \\
&\leq 2 \sum_{k \in S^j} \max\left[\pi_k^j - \mu\left(Q_k\right), 0\right] \\
&\leq 2 \sum_{k \in S^j} \left(\epsilon\beta_k/2\right) = \epsilon \sum_{k \in S^j} \beta_k \leq \epsilon,
\end{aligned}$$

as required.   □

Example A1.7 *We describe the labeling procedure of the previous lemma for the column C in the case where $M = m_l = 14$, $j = j_0 = 3$, $S^3 = \{2,3\}$, and $k = 2$. We assume the $\mathcal{R}$ label of level 1 is $(3,1)$ and the $\mathcal{R}$ label of the top level 14 is $(2,2)$. This situation is drawn in Figure A1.3(a). Note that the number L corresponding to the relatively prime set $\{2,3\}$ is $L = 2$. (For example $2 = 2, 3 = 3, 4 = 2 + 2, 5 = 3 + 2, 6 = 3 + 3, 7 = 2 + 2 + 3$, etc.) We begin by labeling the Top and Bottom blocks in the only way possible, keeping the $\mathcal{R}$ labels. This is shown in the column (b). We then follow the Bottom block with consecutive pairs $(2,1), (2,2)$, until the distance to the lowest level in the Top block (level 13) lies between L and $L+k$, that is, between 2 and 4. This occurs when we have labeled level 9 as $(2,2)$, and this distance is 3 (levels 10, 11, 12). This labeling of the Middle block is shown in (c). The theory ensures that we can fill this Transition block (levels 10, 11, 12) with sequences of $(2,1), (2,2)$ and $(3,1), (3,2), (3,3)$. In fact a single sequence of the latter suffices. The final labeling is shown in (d). Thus the decomposition of* the levels into Bottom, Middle, Transition, and Top blocks is given by

$$\{1,2,3\} \cup \{4,5,6,7,8,9\} \cup \{10,11,12\} \cup \{13,14\}.$$

We can see that regardless of the height M of such a column, the numbers of levels in the Bottom, Top, and Transition blocks are respectively bounded above by $3, 3$, and $L + k = 4$, so at least a fraction $(M - 10)/M$ of the levels are labeled with $(k, \cdot)$, which in this case is $(2, \cdot)$.

We now combine the two lemmas to give a proof of the Multiple Tower Rokhlin Theorem.

Proof We will obtain the required π-skyscraper $\mathcal{P}$ as the limit, with respect to the partition metric, of π-skyscrapers $\mathcal{P}^j$ of type j, obtained using Lemma A1.6. For $j \geq j_0$ choose positive numbers ϵ_j which tend to zero and are sufficiently small so that

$$|d(\mathcal{Q}) - \pi^j| < \epsilon_j \text{ implies } \triangle_j(\mathcal{Q}) < 2^{-(j+1)}$$

for any π-skyscraper $\mathcal{Q}$ of type j. For each $j \geq j_0$ define $s_j = \sum_{k \in S^j} \pi_k$. We recursively construct π-skyscrapers $\mathcal{P}^j$ of type j satisfying

(i) $|d(\mathcal{P}^j) - \pi^j| < \epsilon_j$, $j \geq j_0$, and
(ii) $\|\mathcal{P}^j - \mathcal{P}^{j-1}\| \leq 2^{-j} + \pi_j/s_j$, $j > j_0$.

Since condition (ii) implies that the sequence $\mathcal{P}^j$ is Cauchy, the completeness of the π-skyscrapers with respect to the partition metric ensures that the $\mathcal{P}^j$ converge to a limit π-skyscraper $\mathcal{P}$. Condition (i) then guarantees that $d(\mathcal{P}) = \pi$.

The $\mathcal{P}^j$ are constructed as follows. The first one, $\mathcal{P}^{j_0}$, can be constructed to satisfy the first condition by using the algorithm of Lemma A1.6, with $\alpha = 0$ and $\beta_k = \pi_k^{j_0}$, $k \in S^{j_0}$. Now suppose that $\mathcal{P}^{j_0}, \ldots, \mathcal{P}^{j-1}$ have been found, satisfying both conditions. Since $\mathcal{P}^{j-1}$ satisfies the inequality $|d(\mathcal{P}^{j-1}) - \pi^{j-1}| < \epsilon_{j-1}$, it follows from the choice of ϵ_{j-1} that $\triangle_{j-1}(\mathcal{P}^{j-1}) < 2^{-j}$. Now observe that $\mathcal{P}^{j-1}$ is also of type j, and that

$$\triangle_j(\mathcal{P}^{j-1}) \leq \triangle_{j-1}(\mathcal{P}^{j-1}) + \pi_j/s_j \leq 2^{-j} + \pi_j/s_j.$$

Now apply Lemma A1.6 with $\mathcal{R} = \mathcal{P}^{j-1}$ and $\epsilon = \epsilon_j$ to obtain (as $\mathcal{Q}$) the π-skyscraper $\mathcal{P}^j$ satisfying conditions (i) and (ii). $\qquad\square$

A short proof of Theorem A1.4 for finitely many towers has been obtained by Eigen and Prasad [57] and a short proof of the full version (using an ergodic decomposition theorem) has been obtained by Prikhodcko and Ryzhikov [98].

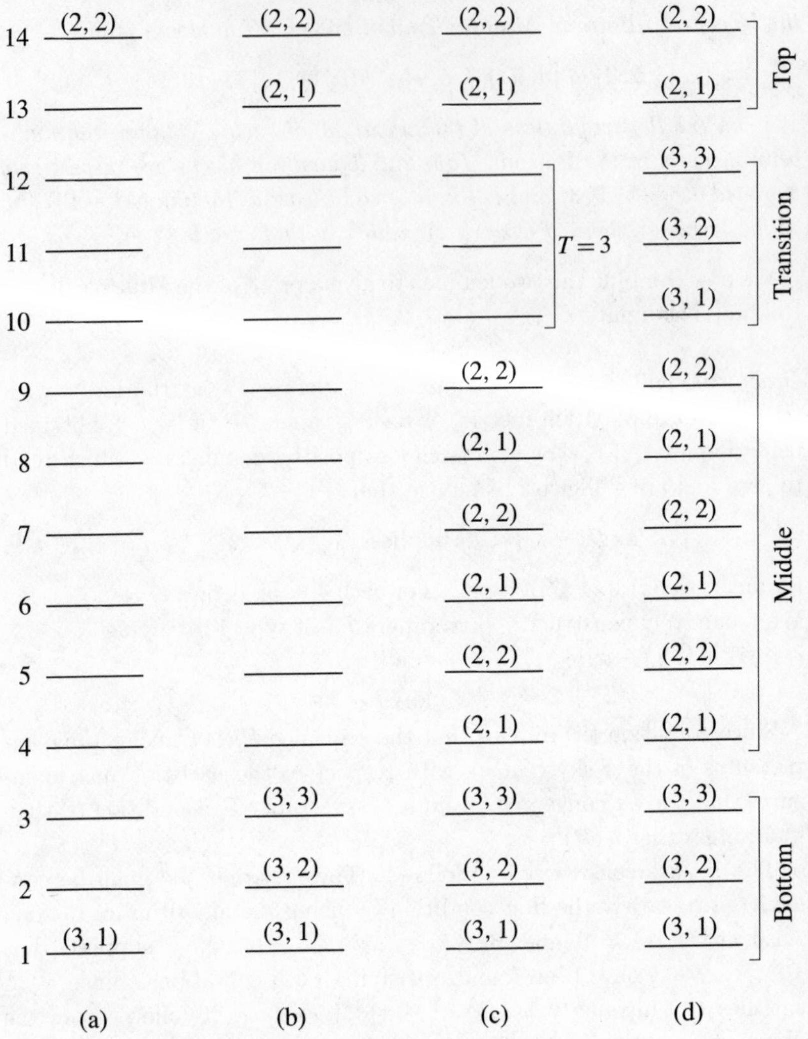

14	(2, 2)	(2, 2)	(2, 2)	(2, 2)	Top
13		(2, 1)	(2, 1)	(2, 1)	
12				(3, 3)	Transition
11			$T = 3$	(3, 2)	
10				(3, 1)	
9			(2, 2)	(2, 2)	Middle
8			(2, 1)	(2, 1)	
7			(2, 2)	(2, 2)	
6			(2, 1)	(2, 1)	
5			(2, 2)	(2, 2)	
4			(2, 1)	(2, 1)	
3		(3, 3)	(3, 3)	(3, 3)	Bottom
2		(3, 2)	(3, 2)	(3, 2)	
1	(3, 1)	(3, 1)	(3, 1)	(3, 1)	
	(a)	(b)	(c)	(d)	

Fig. A1.3. Definition of $\mathcal{Q}$ on D_2

A1.4 Pointwise Conjugacy Approximation

For Lebesgue probability spaces (X, μ), a classical result of Halmos says that the conjugacy class $\mathcal{C}(g) = \{f^{-1}gf : f \in \mathcal{G}[X, \mu]\}$ of any anti-periodic automorphism $g \in \mathcal{G}[X, \mu]$ is dense in $\mathcal{G}[X, \mu]$ in the weak topology (see [72], p. 77). In this section we use the Multiple Tower Rokhlin Theorem to prove a stronger approximation property for $\mathcal{C}(g)$.

We give conditions on automorphisms $h \in \mathcal{G}[X,\mu]$ and measurable sets F such that some conjugate $\hat{g}$ of g *equals* h pointwise a.e. on the set F. Of course we need to assume that $\mu(F) < 1 = \mu(X)$, since otherwise this could only be true if h itself was a conjugate of g. Suppose, for example, that for some subset F_0 of F of measure $1/2$ we have that $h(F_0) = X - F_0$. If a conjugate of g equals h on F then this conjugate, and hence also g, has a periodic set of period 2. Since this is not a property of an arbitrary antiperiodic automorphism, no conjugate of a g without this property can equal h on F. For this reason we must rule out the possibility that h has a nontrivial periodic set. The condition that h is setwise antiperiodic (no nontrivial periodic set) is equivalent to saying that h^m is ergodic for all $m \geq 1$, since a nontrivial periodic set for h of period m is a nontrivial invariant set for h^m. Such an automorphism, for which all positive integer iterates are ergodic, is called *totally ergodic*. The following strengthening of Halmos's result is taken from [13].

Theorem A1.8 *Let* $h, g \in \mathcal{G}[X,\mu]$ *be given, with* g *antiperiodic. Let* F *be a measurable set with* $\mu(F) < 1 = \mu(X)$. *Assume either*

 (i) h *is ergodic and* $\mu(F \cup h(F)) < 1$ *or*
 (ii) h *is totally ergodic.*

Then there is an $f \in \mathcal{G}[X,\mu]$ *such that the conjugate* $\hat{g} = f^{-1}gf$ *of* g *satisfies*

$$\hat{g}(x) = h(x) \text{ for a.e. } x \in F.$$

Proof Consider for each $k = 2, 3, \ldots,$ the subset $A_{k,1} \subset F - h(F)$ consisting of points of F which first leave F on the $(k-1)$th iterate under h. Let $A_{k,i} = h^{i-1}(F_{k,1})$ for $i = 1, 2, \ldots, k$. Note that $A_{k,i} \subset F$ for each $i = 1, 2, \ldots, k-1$, and $A_{k,k} \subset h(F)$. Set $A_{1,1} = X - (F \cup hF)$. Since h is ergodic and $\mu(F) < 1$, it follows that the h-orbit of μ-a.e. point of F eventually leaves F and $\mathcal{A} = \{A_{k,i} : k = 1, \ldots, \infty; \; i = 1, \ldots, k\}$, is a skyscraper partition for h. Furthermore, $\{A_{k,i} : k = 2, 3, \ldots; \; i = 1, 2, \ldots, k\}$ is a partition of the set

$$F \cup h(F) = \bigcup_{k=2}^{\infty} \bigcup_{i=1}^{k} A_{k,i}.$$

The distribution of the measures of the columns of $\mathcal{A}$ is given by $\pi = (\pi_1, \pi_2, \ldots)$ where $\pi_k = \mu\left(\bigcup_{i=1}^{k} A_{k,i}\right)$ for $k = 1, 2, \ldots$ is the measure of the column of height k.

We would like to use the Multiple Tower Rokhlin Theorem (Theorem A1.4) to construct a π-skyscraper for the antiperiodic automorphism g, that is, one with the same size and shape as the skyscraper $\mathcal{A} = \{A_{k,i}\}$ for h. To apply that result we must first check that the ks for which $\pi_k > 0$ are relatively prime. We demonstrate this fact separately for the two alternative assumptions. If we assume (i) that $\mu\left(F \cup h(F)\right) < 1$, then the set $A_{1,1}$ has positive measure, so that $\pi_1 > 0$. Next suppose we are in case (ii) but not case (i) so that h is setwise antiperiodic and $\mu\left(F \cup h(F)\right) = 1$. Suppose in this case that the greatest common divisor of the k's for which $\pi_k > 0$ is some integer $p > 1$. This means that all the column heights of the skyscraper $\{A_{k,i}\}$ are multiples of p. Let

$$D = \bigcup_{j=1}^{\infty} \bigcup_{i=0}^{j-1} A_{pj,ip+1}$$

be the set consisting of the base of the skyscraper $\{A_{k,i}\}$, and every pth level thereafter. Then $\mu(D) = 1/p$ and $h^p(D) = D$, which violates the assumed setwise antiperiodicity of h. Thus we have shown that the column distribution π satisfies the hypothesis of the Multiple Tower Rokhlin Theorem A1.4. Let $\{P_{k,i}\}$ denote the π-skyscraper given by that theorem for the antiperiodic automorphism g. Define $f \in \mathcal{G}[X, \mu]$ on $\bigcup_k A_{k,1}$ so that $f\left(A_{k,1}\right) = P_{k,1}$ for all k – note that this is possible since for each k, both of these sets ($P_{k,1}$ and $A_{k,1}$) have the same measure. Extend f to the rest of X by defining

$$f(x) = g^{i-1} f h^{1-i}(x), \text{ for } x \in A_{k,i}, \ i > 1.$$

It follows from this construction that

$$h(x) = f^{-1} g f(x), \text{ for } x \in A_{k,i}, \ i < k.$$

But since

$$F \subset \bigcup_{k=1}^{\infty} \bigcup_{i=1}^{k-1} A_{k,i},$$

we are done. $\square$

As an illustration of the power of the above type of approximation, in particular the possibility of specifying in advance the set F where the approximation is exact, we use it to give a short proof of the Conjugacy Approximation Theorem, Theorem 8.4. Recall the statement:

Theorem 8.4 *Given any antiperiodic automorphism g in $\mathcal{G}[I^n, \lambda]$, $n \geq$ 2, any volume preserving homeomorphism in $\mathcal{M}[I^n, \lambda]$, and $\epsilon > 0$, there is an f in $\mathcal{G}[I^n, \lambda]$ satisfying $|f^{-1}gf(x) - h(x)| < \epsilon$, for μ-a.e. x.*

Proof We may assume that h is ergodic, since Theorem 7.1 showed that the ergodics are dense $\mathcal{M}[I^n, \lambda]$ in the uniform topology. By Brouwer's Theorem, there is a fixed point $p = h(p)$. Let B be the open $\epsilon/2$-ball centered at p. Define $F = h^{-1}(I^n - B)$. Observe that the closed set $F \cup h(F)$ does not contain p, and consequently has volume less than 1. According to the above theorem, there is an $f \in \mathcal{G}[I^n, \lambda]$ for which $f^{-1}gf = h$ on the set F. For points $y \notin F$, both $f^{-1}gf(y)$ and $h(y)$ belong to B, and consequently are at most a distance ϵ apart. $\qquad\square$

The above proof, which uses condition (i) of Theorem A1.8, can only be used when the underlying manifold has the fixed point property. However, for general manifolds weak mixing homeomorphisms (which are totally ergodic) are generic (see [76] and [7]). Hence in the above proof we can begin by assuming, without loss of generality, that the given homeomorphism h is weak mixing and in particular totally ergodic. If we simply take the set B to be any $\epsilon/2$-ball, the set $F \cup h(F)$ might have measure 1, but this does not matter since now condition (ii) of Theorem A1.8 applies. This approach gives a more general proof, but requires the intermediate result of generic weak mixing in $\mathcal{M}[X, \mu]$.

A1.5 Specified Transition Probabilities

Given an antiperiodic automorphism $g \in \mathcal{G}[X, \mu]$, can we find a partition $\{P_i\}$ of the underlying Lebesgue probability space (X, μ) such that the transition probabilities $\mu(gP_i \cap P_j)/\mu(P_i)$ are equal to given probabilities p_{ij}? We shall consider this question when the index set for the partition is a finite set $S = S_m = \{1, \ldots, m\}$ and also when it is the countable set of natural numbers, denoted $S = S_\infty$, in which case we shall consider that $m = \infty$. In either case, since $\sum_{j \in S} \mu(gP_i \cap P_j) = \mu(P_i)$, we need that $\sum_{j \in S} p_{ij} = 1$.

Such a matrix $\mathbf{P} = (p_{ij})$ will be called *stochastic*. For the reader's convenience we digress in this paragraph to give some terminology from the theory of Markov chains – all of this material can be found in [74]. The matrix $\mathbf{P}$ is called *irreducible* if for every pair of states $i, j \in S$, there is some natural number l such that $p_{ij}^{(l)} > 0$ (i.e., there is a positive probability of visiting j from i in l steps). The *period of a state i* is defined to be $per(i) = \gcd\{l : p_{ii}^{(l)} > 0\}$. When $\mathbf{P}$ is irreducible the period of every state is the same and is called the *period of the matrix* $\mathbf{P}$. If every state has period 1, the matrix is called *aperiodic*. A state i is called *recurrent* if the Markov chain with matrix $\mathbf{P}$ starting at i returns (eventually) to i with probability 1. A recurrent state i is called *positive recurrent* if starting at i the expected return time to i is finite. The matrix $\mathbf{P}$ is called *positive recurrent* if every state is positive recurrent.

In the case of finite m, we shall call the matrix $\mathbf{P} = (p_{ij})$ *mixing* if some power of the matrix has all positive entries (this is equivalent to being both irreducible and aperiodic). In the infinite case we will call it mixing if it is positive recurrent, aperiodic, and irreducible. In either case there is a unique invariant positive distribution vector v_i, $i \in S$, with $v_j = \sum_{i \in S} v_i p_{ij}$, which would have to equal the distribution vector $\mu(P_i)$. As an application of Theorem A1.8, we shall prove the following result originally obtained in [11] for finite S and in [23] for the general case.

Theorem A1.9 *Let g be an antiperiodic automorphism of a Lebesgue probability space (X, Σ, μ) and let $\mathbf{P} = (p_{ij})$, $i, j \in S$ be a mixing stochastic matrix, where S is finite or countable. Then there is a measurable partition $\{P_i\}_{i \in S}$ of (X, Σ) for which*

$$\frac{\mu(gP_i \cap P_j)}{\mu(P_j)} = p_{ij}, \text{ for all } i, j \in S.$$

Proof We begin by recalling the well known construction by which the stochastic matrix (or Markov chain) (p_{ij}) induces a measure preserving transformation. Let X be the product of a countable number of copies of the discrete measure space (S, v), where v is the unique invariant vector described above for $\mathbf{P} = (p_{ij})$. That is, X consists of two-sided infinite sequences $x = (\ldots, x_{-1}, x_0, x_1, \ldots)$, with all the x_k taken from the index or symbol space S. Define a *word* on S to be a finite sequence $w = [w_1, w_2, \ldots, w_{L(w)}]$, where $L(w)$ is called the length of w. For any word w, let X_w be the cylinder set given by $\{x : x_k = w_k, k = 1, \ldots, L(w)\}$. The product measure on X is defined for every word w in S by

$$\mu(X_w) = v_{w(1)} \prod_{k=1}^{L(w)-1} p_{w_k w_{k+1}}.$$

The measure μ is invariant for the left-shift transformation h defined by

$(h(x))_k = x_{k+1}$. The assumption that (p_{ij}) is a mixing matrix suffices to ensure that the shift automorphism h of X is totally ergodic (in fact it is moreover a mixing automorphism – see for example [54, Chapter 8]).

We now consider the antiperiodic automorphism g given in the statement of the theorem. Of course its underlying Lebesgue probability space is not necessarily the product space of (S, v). But since all such spaces are isomorphic it will save us a symbol for the isomorphism if we make this assumption.

Now apply Theorem A1.8 to the totally ergodic shift automorphism h, the antiperiodic automorphism g, and the set $F = \{x : x_2 \neq s\}$, i.e.,

$$F = h^{-1}\left(X - X_{[s]}\right) = \bigcup_{j \neq s} h^{-1}\left(X_{[j]}\right),$$

for a fixed element $s \in S$, where $X_{[j]} = \{x : x_1 = j\}$. (Since $v_s > 0$, we have $\mu(F) < 1$.) Let $\hat{g} = f^{-1}gf$ be the conjugate of g given by Theorem A1.8 which equals h on the set F. Since for any $j \neq s$, $h^{-1}\left(X_{[j]}\right) \subset F$, we have

$$\hat{g}\left(h^{-1}\left(X_{[j]}\right)\right) = h\left(h^{-1}\left(X_{[j]}\right)\right) = X_{[j]}, \text{ or}$$

$$h^{-1}\left(X_{[j]}\right) = \hat{g}^{-1}\left(X_{[j]}\right), \text{ for all } j \neq s, \tag{A1.3}$$

and hence (A1.3) is also true for $j = s$, since $X_{[s]} = X - \bigcup_{j \neq s} X_{[j]}$.

It follows that for all j,

$$\begin{aligned}
p_{ij} &= \mu\left(X_{[i]} \cap h^{-1}\left(X_{[j]}\right)\right) \\
&= \mu\left(X_{[i]} \cap \hat{g}^{-1}\left(X_{[j]}\right)\right) \\
&= \mu\left(\hat{g}\left(X_{[i]}\right) \cap X_{[j]}\right) \\
&= \mu\left(f^{-1}gf\left(X_{[i]}\right) \cap X_{[j]}\right) \\
&= \mu\left(g\left(f\left(X_{[i]}\right)\right) \cap f\left(X_{[j]}\right)\right).
\end{aligned}$$

Thus we see that setting $P_i = f\left(X_{[i]}\right)$ gives the partition required by the theorem. $\qquad\square$

A1.6 Setwise Conjugacy Approximation

Earlier in this chapter, we considered the problem of approximating a given automorphism $h \in \mathcal{G}[X, \mu]$ by some conjugate $\hat{g}$ of a given antiperiodic automorphism $g \in \mathcal{G}[X, \mu]$ in the sense that $\hat{g}$ equals h *pointwise* on a given set. We found that in general we required that h be totally ergodic (setwise aperiodic) for such an approximation to be possible.

We now consider a weaker type of approximation, where $\hat{g}$ must equal h *setwise* on a finite collection of sets A_k, $k = 1, \ldots, l$, which will be possible with a weaker hypothesis on the target automorphism h. Note that if $\hat{g}(A_k) = h(A_k)$, $k = 1, \ldots, l$, then it would follow that $\hat{g}(A) = h(A)$ whenever A belongs to the subalgebra Λ of Σ generated by the sets A_k. Now suppose that the set map $h|\Lambda$ has a nontrivial periodic point, that is, a set $A \in \Lambda - \{\emptyset, X\}$ such that $h^i(A) \in \Lambda$, $i = 1, \ldots, k$, and $h^k(A) = A$. Then if the map $\hat{g}$ equaled h setwise on Λ, it would also have a nontrivial periodic set. This is not possible if g has no such periodic set. So we must exclude this possibility for h. Note that we can still allow h to have a periodic set, but not all of its iterates can belong to Λ. The possibility of setwise approximation with this assumption is stated below in Theorem A1.11. The need for such a result comes mainly from requirements concerning manifolds with ends, where the ends must be preserved exactly in any approximation.

Before getting to Theorem A1.11, we must adapt the results of the previous section. Let $\mathbf{T} = (t_{ij})$ be an $m \times m$ mixing matrix with entries in $\{0, 1\}$. Let $\hat{T} : \Gamma \to \Gamma$ denote the set map induced by $\mathbf{T}$ on the power set Γ of $\{1, 2, \ldots, m\}$, that is, $j \in \hat{T}(\gamma)$ if and only if $i \in \gamma$ for some i with $t_{ij} = 1$. We call a set $\gamma \in \Gamma$ *critical* if $t_{ij} = 1$ and $j \in \hat{T}(\gamma)$ imply $i \in \gamma$, and denote by Γ_1 the subalgebra of Γ consisting of all the critical sets. We say that a probability distribution $v = (v_1, \ldots, v_m)$ is *consistent* with $\mathbf{T}$ if it is the unique invariant distribution for some (necessarily mixing) stochastic matrix (p_{ij}) where $p_{ij} = 0$ if and only if $t_{ij} = 0$. It is easily verified that v is consistent with the 0–1 matrix $\mathbf{T}$ if and only if it satisfies both of the following conditions.

(1) $\sum_{i \in \gamma} v_i = \sum_{j \in \hat{T}(\gamma)} v_j$, for all $\gamma \in \Gamma_1$; and

(2) $\sum_{i \in \gamma} v_i < \sum_{j \in \hat{T}(\gamma)} v_j$, for all $\gamma \in \Gamma - \Gamma_1$.

Theorem A1.10 *Let* $\{P_i\}_{i=1}^m$ *be a partition of a finite Lebesgue space* (X, Σ, μ) *whose distribution* $(\mu(P_1), \ldots, \mu(P_m))$ *is consistent with a given* $m \times m$, *0–1 mixing matrix* $\mathbf{T}$. *Then given any antiperiodic automorphism* $g \in \mathcal{G}[X, \mu]$, *there is a conjugate* $\hat{g} = f^{-1}gf$, $f \in \mathcal{G}[X, \mu]$, *such that* $\mu(\hat{g}P_i \cap P_j) = 0$ *if* $t_{ij} = 0$.

Proof By definition of consistency, the distribution $(\mu(P_1), \ldots, \mu(P_m))$ is invariant with respect to some stochastic matrix which has the same signs as $\mathbf{T}$, and consequently is mixing. By Theorem A1.9 there is a partition $\{Q_i\}_{i=1}^m$ with the same distribution as $\{P_i\}_{i=1}^m$ and $\mu(gQ_i \cap Q_j) =$

0 if $t_{ij} = 0$. Let $f \in \mathcal{G}[X, \mu]$ be an automorphism with $P_i = f^{-1}Q_i$, for $i = 1, \ldots, m$. Then $\hat{g} = f^{-1}gf$ is the required conjugate. $\square$

We can now state and prove our result on setwise conjugacy approximation.

Theorem A1.11 *Let $h, g \in \mathcal{G}[X, \Sigma, \mu]$, where $\mu(X) = 1$ and g is an antiperiodic automorphism. Let Λ be a finite subalgebra of Σ such that $h|\Lambda$ has no nontrivial periodic set (this means that there is no set $A \in \Lambda - \{X, \emptyset\}$ such that for some k, $h^i(A) \in \Lambda$ for $i = 1, \ldots, k$ and $h^k(A) = A$). Then there is some conjugate $\hat{g} = f^{-1}gf$, $f \in \mathcal{G}[X, \mu]$, such that $\hat{g}(A) = h(A)$ for all $A \in \Lambda$.*

Proof Let $A_k, k = 1, \ldots, l$, denote the atoms of Λ. Let $\{P_i\}_{i=1}^m$ be a measurable partition of X which refines the partitions of X determined by the atoms of Λ and by the atoms of $h(\Lambda)$. Define an $m \times m$, 0–1 matrix $\mathbf{T} = (t_{ij})$ by

$$t_{ij} = \begin{cases} 1, & \text{if } P_i \subset A_k \text{ and } P_j \subset h(A_k), \text{ for some } k, \\ 0, & \text{otherwise.} \end{cases}$$

Observe that $\gamma \subset \{1, \ldots, m\}$ is a critical set for $\mathbf{T}$ (i.e., $\gamma \in \Gamma_1$) if and only if $\bigcup_{i \in \gamma} P_i \in \Lambda$, and that the distribution $(\mu(P_1), \ldots, \mu(P_m))$ is consequently consistent with $\mathbf{T}$. To show that $\mathbf{T}$ is a mixing matrix it is sufficient to prove that, for any nonempty set $\gamma \in \Gamma$, the sequence

$$\mu(\gamma), \mu(\hat{T}(\gamma)), \mu(\hat{T}^2(\gamma)), \ldots$$

is eventually equal to $\mu(X)$, where $\mu(\gamma) = \sum_{i \in \gamma} \mu(P_i)$. (If N is the longest number of steps ever required to reach 1, then $\mathbf{T}^N > 0$.) It follows from the definition of t_{ij} that $\mu(\gamma) \leq \mu(\hat{T}(\gamma))$, with equality only for critical sets $\gamma \in \Gamma_1$. Since the algebra Λ is assumed to be finite, it follows that the only way the above nondecreasing sequence can fail to reach $\mu(X)$, is if $\hat{T}|\Gamma_1$ has a nontrivial periodic point γ_0. But then the set $A = \bigcup_{i \in \gamma_0} P_i$ would be a nontrivial periodic point of $h|\Lambda$. But since this possibility has been excluded by assumption, our argument shows that $\mathbf{T}$ is indeed a mixing matrix. We now apply Theorem A1.10 and observe that the conjugate $\hat{g}$ of g it gives us satisfies the requirements of the present theorem. $\square$

The above result on setwise approximation is of particular value when there is a metric on the underlying space X. The following corollary illustrates the nature of the applications. The first part of the following result establishes the Setwise Conjugacy Approximation Theorem for both I^n (Theorem 8.1) and an arbitrary compact manifold (Theorem 10.1). (Recall that a metric space is called totally bounded if for every positive ϵ it can be covered by finitely many ϵ-balls.) The following result is restated in Chapter 16 in a notation better suited to the applications in that section.

Corollary A1.12 (Setwise Conjugacy Approximation) *Let ρ be a totally bounded, connected metric on a Lebesgue probability space (X, Σ, μ) such that every nonempty open set has positive μ-measure. Let $h, g \in \mathcal{G}[X, \mu]$, with g antiperiodic. Then given any $\epsilon > 0$ there is a conjugate $\hat{g} = f^{-1}gf$, for some $f \in \mathcal{G}[X, \mu]$, such that $\rho\left(\hat{g}(x), h(x)\right) < \epsilon$, for μ-a.e. $x \in X$.*

Furthermore suppose Λ is a finite subalgebra of Σ such that $h|\Lambda$ has no nontrivial periodic set, and let C be the union of all atoms of Λ whose image under h is connected and relatively compact. Then given any $\epsilon > 0$ there is a conjugate $\hat{g} = f^{-1}gf$, for some $f \in \mathcal{G}[X, \mu]$, such that $\rho\left(\hat{g}(x), h(x)\right) < \epsilon$ for μ-a.e. $x \in C$, and $\hat{g}(A) = h(A)$ for all $A \in \Lambda$.

Proof As the first paragraph of the corollary is a special case of the second with $\Lambda = \{\emptyset, X\}$ and $C = X$, we will only prove the second.

As the proof of the second paragraph is very similar to that of the previous theorem, we emphasize only the differences. Let A_k, $k = 1, \ldots, l$, denote the atoms of Λ, and assume that $h(A_k)$ is connected for all $k \leq L$, for some $L \leq l$. Let $\{P_i\}_{i=1}^m$ be a measurable partition of X which refines the partitions of X determined by the atoms of Λ and by the atoms of $h(\Lambda)$, and in addition satisfies

$$\rho\left(P_i\right) < \epsilon/2 \text{ and } \rho\left(hP_i\right) < \epsilon/2, \ i = 1, \ldots, m,$$

where $\rho(S)$ denotes the diameter of a set S. Such a partition exists because the metric space (X, ρ) is assumed to be totally bounded. Define the $m \times m$, 0–1 matrix $\mathbf{T} = (t_{ij})$ as follows (where $\bar{S}$ denotes the closure of a set S):

$$t_{ij} = \begin{cases} 1, & \text{if } P_i \subset A_k, P_j \subset h(A_k), \text{ for some } k > L \\ 1, & \text{if } P_i \subset A_k, P_j \subset h(A_k), \overline{h(P_i)} \cap \overline{P_j} \neq \emptyset, \ k \leq L \\ 0, & \text{otherwise.} \end{cases}$$

The proof that $\mathbf{T}$ is mixing and that the distribution $(\mu(P_1), \ldots, \mu(P_m))$ is consistent with $\mathbf{T}$ is the same as for the previous theorem, except that the connectivity of $h(A_k), k \leq L$, is used to identify Γ_1 with the algebra

Λ. Let $\hat{g}$ denote the conjugate $\hat{g} = f^{-1}gf$, $f \in \mathcal{G}[X, \mu]$, of g, such that $\hat{g}(A) = h(A)$, $\forall A \in \Lambda$, given by Theorem A1.10.

To establish the final metric inequality assume that $P_i \subset C$, or equivalently that $P_i \subset A_k$ for some $k \leq L$. In this case we have that the diameters of the sets below satisfy

$$\rho(\hat{g}P_i \cup hP_i) \leq \max_{\{j:t_{ij}=1\}} \rho(P_j \cup hP_i) \leq \max_{\{j:t_{ij}=1\}} [\rho(P_j) + \rho(hP_i)] < \epsilon,$$

because $\overline{h(P_i)} \cap \overline{P_j} \neq \emptyset$. □

For some applications, we will need only the following special case of the previous result. We have already shown that any automorphism of a finite measure space can be uniformly approximated by an ergodic one. The next result says that this can be accomplished without changing the image of a given set A, as long as the set A itself is not invariant. For reasons of intended application, the theorem is expressed in terms of an initial transformation which is only defined on the set A.

Corollary A1.13 *Let ρ be a totally bounded, connected metric on a Lebesgue probability space (X, Σ, μ) such that every nonempty open set has positive measure. Let A be a measurable set and let $\check{h} : A \to X$ be any μ-preserving injection, where the domain A is not $\check{h}$-invariant $(\mu(\check{h}(A) \bigtriangleup A) > 0)$ and $\check{h}(A)$ is connected. Then given any $\epsilon > 0$, there is μ-preserving injection $\check{g} : A \to \check{h}(A)$ such that $\rho(\check{g}(x), \check{h}(x)) < \epsilon$ for μ-a.e. $x \in A$, and there are no nontrivial $\check{g}$-invariant subsets of A.*

Proof Without loss of generality we may assume that $\mu(A \cap \check{h}(A)) \neq 0$, for otherwise we may simply take $\check{g} = \check{h}$. Let $h \in \mathcal{G}[X, \mu]$ denote any fixed extension of $\check{h}$, and consider the subalgebra of Σ given by $\Lambda = \{\emptyset, X, A, X - A\}$. Then the hypotheses of the previous theorem hold, with either $C = A$ or X, but certainly $C \supseteq A$. Let $g \in \mathcal{G}[X, \mu]$ be any ergodic automorphism, and let $\hat{g} \in \mathcal{G}[X, \mu]$ be the conjugate of g which approximates h in the sense of the previous theorem. Consequently $h(A) = \hat{g}(A)$ and hence the restriction of $\hat{g}$ to A, denoted $\check{g}$, satisfies the requirements of the theorem. □

A1.7 Infinite Measure Constructions

Most of the constructions up to now in this chapter have been performed on a Lebesgue probability space. For some applications to noncompact manifolds, we will also need to construct towers on spaces of infinite measure. In this section we obtain tower constructions and conjugacy results for spaces of infinite measure, although somewhat stronger assumptions on the shape of the towers are required.

In this section we assume that the underlying measure space (X, Σ, μ) is an infinite, sigma finite, nonatomic Lebesgue space. Recall that an automorphism of such a space is said to be ergodic if an invariant set has measure zero or its complement has measure zero. Before stating the main results of this section, we begin by proving a simple lemma on which the main results are based. All the results of this section are taken from a joint paper with Jal Choksi [18].

Lemma A1.14 *Let g be an ergodic automorphism of the infinite Lebesgue space (X, Σ, μ) and let m be a given positive integer. For each measurable set S with $\mu(S) < \infty$, and extended nonnegative real number r $(0 \leq r \leq \infty)$, there is a measurable set R such that $\mu(R) = r$ and the sets $S, R, g(R), \ldots, g^{m-1}(R)$ are pairwise disjoint.*

Proof We may assume without loss of generality that $r = \infty$, since a set of infinite measure has subsets of arbitrary finite measure. Define $S_0 = S$ and $S_k = gS_{k-1} - S_0$, recursively for $k = 1, 2, \ldots$. The sets S_k are disjoint by construction, and it follows from the assumed ergodicity of g that $\mu(S_k)$ converges monotonically to zero and $X = \bigcup_{k=0}^{\infty} S_k$. Using the given integer m, we define the required set R by

$$R = \bigcup_{i=1}^{\infty} g^{1-m} S_{im}.$$

Note R consists of points in $X - S$ whose next $m - 1$ iterates do not enter S. From this it is clear that $S, R, g(R), \ldots, g^{m-1}(R)$ are pairwise disjoint, as required. Furthermore, since

$$\mu\left(X - [S \cup R \cup g(R) \cup \cdots \cup g^{m-1}(R)]\right)$$
$$\leq \quad \mu\left(S \cup g^{-1}S \cup \cdots \cup g^{1-m}(S)\right) \leq m\,\mu(S) < \infty,$$

it follows that R must have infinite measure. $\qquad \square$

We now establish an infinite measure form of the Multiple Tower Rokhlin Theorem, which replaces the 'relatively prime heights' condition by the stronger assumption that the column of height 1 has infinite measure. Note that the former assumption of aperiodicity has been strengthened to ergodicity.

Theorem A1.15 *Let g be an ergodic automorphism of an infinite, sigma finite nonatomic Lebesgue space (X, Σ, μ). Let $(p_1, p_2, \ldots)$ be a sequence of nonnegative extended real numbers $(0 \leq p_k \leq \infty)$ with $p_1 = \infty$. Then there is a skyscraper partition $\{E_{k,i}\}$, $k = 1, \ldots, \infty$, $i = 1, \ldots, k$, of X satisfying*

$$E_{k,i} = g^{i-1} E_{k,1} \text{ and}$$
$$\mu(E_{k,1}) = p_k.$$

Proof We first deal with the case where the p_k are all finite for $k > 1$. Apply the above lemma with $S = S_2 = \emptyset$, $r = r_2 = p_2$, and $m = m_2 = 2$, to obtain a set R_2 with $\mu(R_2) = r_2$ such that S_2, R_2, gR_2 are pairwise disjoint. Proceed recursively as follows. For $k \geq 3$, let B_{k-1} be any set of measure 1 which is disjoint from the finite measure set

$$S_{k-1}^* = S_{k-1} \cup R_{k-1} \cup g(R_{k-1}) \cup \cdots \cup g^{k-2}(R_{k-1}).$$

Apply the above lemma to the set $S = S_k = S_{k-1}^* \cup B_{k-1}$, with $r = r_k = p_k$ and $m = m_k = k$, to obtain a set R_k with $\mu(R_k) = p_k$ and $S_k, R_k,$ $g(R_k), \ldots, g^{k-1}(R_k)$ pairwise disjoint. For $k > 1$ and $1 \leq i \leq k$, define $E_{k,i} = g^{i-1}(R_k)$, and set

$$E_{1,1} = X - \bigcup_{k=2}^{\infty} \bigcup_{i=1}^{k} E_{k,i}.$$

Note that since $E_{1,1}$ contains all the disjoint sets $B_2, B_3, \ldots$, which are assumed to each have measure 1, $E_{1,1}$ must have infinite measure. This completes the proof for the case where the p_k are all finite for $k > 1$.

For the general case where any of the p_k may be infinite, let (m_j, r_j) be a sequence of pairs where m_j is an integer greater than 1 and r_j is a finite positive number such that for each $k \geq 2$,

$$\sum_{\{j : m_j = k\}} r_j = p_k.$$

As for the proof of the special case, use the lemma above to recursively construct sets R_j with $\mu(R_j) = r_j$ and $S_j, R_j, g(R_j), \ldots, g^{m_j-1}(R_j)$ pairwise disjoint. As above, $S_j = S_{j-1}^* \cup B_{j-1}$, where S_{j-1}^* is the union of the previously defined R 's and S 's, and B_{j-1} is a set of measure 1 which is disjoint from S_{j-1}^*. For $k \geq 2$, define

$$E_{k,1} = \bigcup_{\{j : m_j = k\}} R_j, \text{ and } E_{k,i} = g^{i-1} E_{k,1} \text{ for } i \leq k.$$

As before for the special case, we complete the proof by defining

$$E_{1,1} = X - \bigcup_{k=2}^{\infty} \bigcup_{i=1}^{k} E_{k,i}.$$

$\square$

As a corollary of this result, we obtain a pointwise conjugacy approximation theorem for the infinite measure case. This generalizes the similar theorem of Choksi and Kakutani ([50], Theorem 6) in which the set F was assumed to have finite measure. In our applications, we will need to be able to consider cases where F has infinite measure.

Corollary A1.16 *Let $\mathcal{G}[X, \mu]$ denote the group of all automorphisms of the infinite sigma finite nonatomic Lebesgue space (X, Σ, μ). Let h and g be two ergodic automorphisms in $\mathcal{G}[X, \mu]$. Let F be any measurable subset of X such that*

$$\mu\left(X - (F \cup hF)\right) = \infty.$$

Then there is an automorphism $f \in \mathcal{G}[X, \mu]$ such that the conjugate $\hat{g} = f^{-1}gf$ of g satisfies

$$\hat{g}(x) = h(x) \text{ for } \mu\text{-a.e. } x \in F.$$

Proof The proof is essentially the same as the finite measure proof (Theorem A1.8) in this chapter. We begin by writing

$$A_{1,1} = X - (F \cup hF) \text{ and } F \cup hF = \bigcup_{k=2}^{\infty} \bigcup_{i=1}^{k} A_{k,i},$$

where for $k > 1$, $A_{k,i} = h^{i-1}A_{k,1}$ and $A_{k,1}$ is the subset of F consisting of points which first leave F on the $(k-1)$th iterate of h. Define $p_k = \mu(A_{k,1})$, $k \geq 1$, and apply the previous theorem to the sequence p_k and the ergodic automorphism g to produce the skyscraper partition $P_{k,i}$ for g. By construction, we have $\mu(P_{k,i}) = \mu(A_{k,i})$, so we may define $f \in \mathcal{G}[X, \mu]$ on the set $\cup_k A_{k,1}$ so that $f(A_{k,1}) = P_{k,1}$. Extend f to all of X by defining

$$f(x) = g^{i-1}fh^{1-i}(x) \text{ for } x \in A_{k,i}.$$

It follows that

$$h(x) = f^{-1}gf(x) \text{ for } x \in F \subset \bigcup_{k=2}^{\infty} \bigcup_{i=1}^{k-1} A_{k,i}.$$

$\square$

It is likely that the results of this infinite measure section can be improved. The versions given here are however sufficient for our applications to homeomorphisms of noncompact manifolds. Corollary A1.16 is applied in Chapter 17 where it is restated as Theorem 17.2.

A related conjugacy result for the infinite measure case is the following theorem obtained by the authors, which is used in Chapter 17. The proof is long and can be found in [22].

Theorem A1.17 *Let h, $g \in \mathcal{G}[X, \mu]$ be ergodic automorphisms of the infinite sigma finite nonatomic Lebesgue space (X, Σ, μ). Let $X = X_0 \cup X_1 \cup \cdots \cup X_m$ be a measurable partition with $0 < \mu(X_0) < \infty$ and $\mu(X_i) = \infty$ for $i = 1, \ldots, m$. Assume that the $m \times m$ 0-1 matrix $\mathbf{T}$, defined for $i, j > 0$ by $t_{ij} = 1$ if and only if $\mu(hX_i \cap X_j) = \infty$, is mixing. Then there is an automorphism $f \in \mathcal{G}[X, \mu]$ such that the conjugate $\hat{g} = f^{-1}gf$ satisfies*

$$\hat{g}(x) = h(x) \text{ for } \mu\text{-a.e. } x \text{ in } X_0, \text{ and}$$
$$\hat{g}(X_i) = h(X_i) \text{ for } i = 1, \ldots, m.$$

The above result is a hybrid pointwise–setwise approximation. It says that there is a conjugate of any given ergodic automorphism which equals h pointwise on a finite measure set, and setwise on a finite family of infinite measure sets. In the application given in Chapter 17, the finite measure set is a compact separating set together with the finite measure components of its complement, and the infinite measure ones are the infinite measure components of its complement.

Appendix 2
Homeomorphic Measures

A2.1 Introduction

This final chapter is devoted to establishing sufficient conditions for two OU measures μ and ν on a sigma compact manifold X to be 'homeomorphic'. Recall that this means there exists a homeomorphism h of X onto itself satisfying $\nu(A) = \mu(h(A))$ for all measurable sets A, which we write simply as $\nu = \mu h$. The general result of this type is due to Berlanga and Epstein [38], who use in their proof the important special case of the theorem for compact manifolds due to Oxtoby and Ulam [88] and von Neumann [105]. In Chapter 9 we developed various corollaries of the compact version of the theorem, which was stated for the unit cube in Theorem 9.1. The version for a general compact manifold is given here as Corollary A2.6, and the version for sigma compact manifolds is given here as Theorem A2.8. In this chapter we outline the original proofs of the compact case and then the sigma compact case. As such, this is the only chapter of the book that is not based on the work of the authors.

Recall that a Borel measure on a manifold is called an OU measure if it is zero for points, positive on open sets, and zero on the boundary set. Since singleton sets, open sets, and the boundary set are all preserved under homeomorphism, it follows that any Borel measure homeomorphic to an OU measure must also be an OU measure. Since the whole manifold X is also preserved under homeomorphism it follows that homeomorphic measures must also have the same total measure ($\mu(X) = \nu(X)$). For compact manifolds, these necessary conditions for measures to be homeomorphic are in fact sufficient, as stated in the Homeomorphic Measures Theorem, Theorem 9.1 for the cube, and Corollary A2.6 for general compact manifolds.

For noncompact manifolds, the conditions mentioned in the previous paragraph may not be sufficient for two measures to be homeomorphic. We require additional conditions on the induced measures on the ends. If the two measures μ and ν are homeomorphic via a homeomorphism h with $\mu = \nu h$ then the ends E_μ^∞ and E_ν^∞ which are infinite with respect to μ and ν respectively are related by the equation $E_\nu^\infty = h^* E_\mu^\infty$. If we require for simplicity that the homeomorphism h is end preserving, then h^* is the identity and hence $E_\nu^\infty = E_\mu^\infty$, or $\mu^* = \nu^*$. Consequently this requirement that both measures induce the same measure on the ends is a necessary condition for ν and μ to be homeomorphic via an end preserving homeomorphism. (Of course in the compact case both sides of the set identity are the empty set, so it is always satisfied.) The theorem of Berlanga and Epstein (Theorem A2.8) says that if this condition is added to those required in the compact case, the resulting set of conditions is necessary and sufficient for ν and μ to be homeomorphic via an end preserving homeomorphism.

Finally we note there are infinite dimensional analogs of the Homeomorphic Measures Theorem on the Hilbert cube, I^∞ (in [92] by Oxtoby and Prasad and also by B. Weiss (unpublished) [106]); the extension of the result to Hilbert cube manifolds is due to Prasad [97].

A2.2 Homeomorphic Measures on the Cube

We devote this section to a proof of the Homeomorphic Measures Theorem (Theorem 9.1) for I^n, the consequences of which we derived in Chapter 9. For the convenience of the reader, we repeat this important result.

Homeomorphic Measures Theorem *A Borel measure μ in the n-cube I^n is homeomorphic to Lebesgue measure λ, i.e., $\mu = \lambda f$ for some homeomorphism f of I^n, if and only if μ satisfies the following four conditions (which characterize OU probability measures):*

(i) *μ is zero for points (i.e., nonatomic)*

(ii) *μ is zero on the boundary*

(iii) *μ is a probability measure*

(iv) *μ is positive on open sets (μ is locally positive).*

Furthermore the homeomorphism f can be chosen to be the identity on the boundary of I^n.

Since n-dimensional volume measure λ satisfies conditions (i)–(iv) and every homeomorphism preserves singleton sets, the boundary and the open sets, it follows that any measure homeomorphic to λ obviously satisfies these four conditions. The proof that these four conditions on μ suffice for μ to be homeomorphic to λ will follow from a series of lemmas which will form a large part of this section. The main idea is to construct h so that $\mu(R) = \lambda h(R)$ for all sets R of a rectangular subdivision of I^n. Then h is modified within each cell R of the rectangular subdivision so as to secure equality within each cell of a finer subdivision of R. With some care, a sequence of modifications will be performed so that the limiting map will be a homeomorphism with the desired properties.

First we prove an easy lemma about Borel measures.

Lemma A2.1 *Let μ be a finite nonatomic Borel measure on I^n that is zero on the boundary, and let $0 \le \alpha < \beta \le \mu(I^n)$. There is an open set V in the interior of I^n such that $\alpha < \mu(V) < \beta$.*

Proof Since the measure space is compact and points have measure zero there is a number $\gamma > 0$ such that $\mu(U) < \beta - \alpha$ for every open set of diameter less than γ. Choose a compact set K in the interior of I^n with $\mu(K) > \alpha$ and let δ be a positive number less than γ such that the δ-neighborhood of K is contained in the interior of the cube I^n. Divide the cube into rectangles of diameter less than δ with sides of μ-measure zero (see for example, the proof of Lemma 9.4). Let $\{U_1, \ldots, U_m\}$ be the interiors of the elements of this partition that meet K. Let $V = U_1 \cup \cdots \cup U_k$, where k is smallest integer such that $\mu(V) > \alpha$. Then $\alpha < \mu(V) < \beta$. $\qquad\qquad\square$

The basic lemma required to prove the Homeomorphic Measures Theorem (Theorem 9.1) is the following:

Lemma A2.2 *Let μ be a finite nonatomic locally positive Borel measure in I^n which is zero on the boundary. Fix some coordinate k between 1 and n, and let $R_1 = \{x \in I^n : x_k \le c\}$ and $R_2 = \{x \in I^n : x_k \ge c\}$ be rectangles intersecting in the section $P = \{x \in I^n : x_k = c\}$. Then for any two positive numbers α_1 and α_2 with $\alpha_1 + \alpha_2 = \mu(I^n)$ there is an $h \in \mathcal{H}[I^n, \partial I^n]$ such that $\mu h(R_i) = \alpha_i$ for $i = 1, 2$.*

Note that what makes this lemma hard is that the image of the section P under a homeomorphism h of I^n (note this image is common to both $h(R_1)$ and $h(R_2)$) could have positive μ-measure. The main part of this

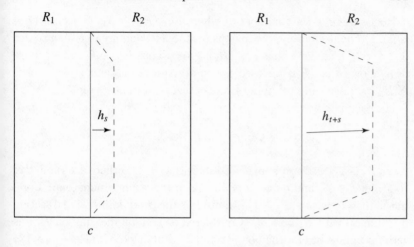

Fig. A2.1. $h_t(R_1)$ includes more of R_2 as t increases. Note that the horizontal axis is the kth coordinate and the other coordinates are represented by the vertical axis.

lemma is to show that such a situation arises for only an exceptional set of homeomorphisms in the sense of Baire category, i.e., a first category set in $\mathcal{H}[I^n, \partial I^n]$. Another approach to this problem is taken by Goffman and Pedrick (see for example [65] or [66]).

Proof Let $\mathcal{F}$ denote the set of all homeomorphisms $h \in \mathcal{H}[I^n, \partial I^n]$ such that $\mu h(R_i) \geq \alpha_i$ $(i = 1, 2)$. We shall establish the existence of the required homeomorphism by a Baire category argument in $\mathcal{F}$. First note that $\mathcal{F}$ is a closed subset of $\mathcal{H}[I^n, \partial I^n]$ since if h_j is any sequence of homeomorphisms in $\mathcal{F}$ converging uniformly to h, then any ϵ-neighborhood of hR_i, $i = 1, 2$ contains $h_j R_i$ for all sufficiently large j and hence has μ-measure at least α_i. This being true for any $\epsilon > 0$, it follows that $\mu h(R_i) \geq \alpha_i$ for $i = 1, 2$, and so $h \in \mathcal{F}$. To show that $\mathcal{F}$ is nonempty note that unless it contains the identity we must have either $\mu(R_1) < \alpha_1$ or $\mu(R_2) < \alpha_2$. We may assume without loss of generality that $\mu(R_1) < \alpha_1$. By deforming I^n by a continuous family of homeomorphisms $h_t \in \mathcal{H}[I^n, \partial I^n]$ $(0 < t < 1)$ in such a way that as t increases, $h_t R_1$ includes more and more of the interior of R_2, we can ensure that (see Figure A2.1)

(i) $\mu h_t(R_1)$ is monotone increasing and continuous on the right
(ii) $\mu h_t(R_1)$ tends to $\mu(R_1)$ as $t \to 0$ and to $\mu(I^n)$ as $t \to 1$.

Hence there is least number t_0 where $\mu h_{t_0}(R_1) \geq \alpha_1$. Since $\mu h_t(R_2)$ is monotone decreasing, continuous on the left, and greater than α_2 for $0 < t < t_0$, it follows that $\mu h_{t_0}(R_2) \geq \alpha_2$. Thus $h_{t_0} \in \mathcal{F}$ and since we have shown that $\mathcal{F}$ is a closed nonempty subset of the complete space $\mathcal{H}[I^n, \partial I^n]$, we can now apply Baire category arguments on $\mathcal{F}$.

For each positive integer j let

$$\mathcal{F}_j = \left\{ h \in \mathcal{F} : \mu h(R_1) \geq \alpha_1 + \frac{1}{j} \right\}.$$

Then $\mathcal{F}_j$ is a closed subset of $\mathcal{F}$. To show that it is nowhere dense relative to $\mathcal{F}$, let $h \in \mathcal{F}_j$ and define $\nu = \mu h$. Then ν is a finite nonatomic Borel measure in I^n with $\nu(P) \geq 1/j$ where P is the $(n-1)$-dimensional section in common with R_1 and R_2. Considered as a Borel measure on P, ν is zero for points and for the boundary of P. Putting $\alpha = \nu(R_1) - \alpha_1 - 1/j$ and $\beta = \nu(R_1) - \alpha_1$, we have $0 \leq \alpha < \beta \leq \nu(P)$. Hence by the previous lemma, applied to I^{n-1} identified with the section P, there is a relatively open subset V of P such that $V \cap P \cap \partial I^n = \emptyset$ and $\alpha < \nu(V) < \beta$. Now let g be a homeomorphism of I^n that fixes the boundary and also the points of $P - V$, but let it displace all points of V slightly into the interior of R_1. (This can be done so that g moves points only in the kth coordinate direction.) We can choose g arbitrarily close to the identity. Since $g(R_2) \supset R_2$, we have $\nu g(R_2) \geq \alpha_2$. Because $\nu(R_1) - \alpha_1 - 1/j < \nu(V) < \nu(R_1) - \alpha_1$, it follows that $\alpha_1 < \nu(R_1 - V) < \alpha_1 + 1/j$, and thus for all sufficiently small g, $\alpha_1 < \nu g(R_1) < \alpha_1 + 1/j$. As $\nu = \mu h$, we therefore have $\alpha_1 < \mu h g(R_1) < \alpha_1 + 1/j$ and $\mu h g(R_2) \geq \alpha_2$. Hence $hg \in \mathcal{F} - \mathcal{F}_j$ for all sufficiently small g, and so $\mathcal{F}_j$ is a nowhere dense subset of $\mathcal{F}$. It follows that the equation $\mu h(R_1) = \alpha_1$ holds for all h in a dense G_δ subset of $\mathcal{F}$, and similarly for the equation $\mu h(R_2) = \alpha_2$. The intersection of these two dense G_δ subsets is also a dense G_δ subset of $\mathcal{F}$. Any h belonging in the intersection of these two sets of homeomorphisms has the required properties. □

The next lemma follows easily by induction from the previous lemma.

Lemma A2.3 *Let μ be any nonatomic finite locally positive Borel measure on I^n which is zero on the boundary, and let $R_1, \ldots, R_N$ be the cells of any rectangular subdivision of I^n, and let $\alpha_1, \ldots, \alpha_N$ be positive numbers with $\alpha_1 + \cdots + \alpha_N = \mu(I^n)$. Then there is a homeomorphism $h \in \mathcal{H}[I^n, \partial I^n]$ such that $\mu h(R_i) = \alpha_i$, for $i = 1, \ldots, N$.*

By a rectangular subdivision $\mathcal{P}$ of I^n, we mean a subdivision of I^n

into rectangles (the subdivision is not a partition because the rectangles may have some sides in common). Let $|\mathcal{P}|$ denote the maximum of the diameters of the members of $\mathcal{P}$.

Lemma A2.4 *Let μ and ν be two OU measures on I^n, and let $\mathcal{P}$ be a rectangular subdivision of I^n such that $\mu(R) = \nu(R)$ and $\mu(\partial R) = \nu(\partial R) = 0$ for each $R \in \mathcal{P}$. For any $\epsilon > 0$ there exists a rectangular subdivision $\mathcal{P}'$ refining $\mathcal{P}$ with $|\mathcal{P}'| < \epsilon$, and an $h \in \mathcal{H}[I^n, \partial I^n]$ that moves points only in the interior of each $R \in \mathcal{P}$, such that $\nu(R') = \mu h(R')$ for each $R' \in \mathcal{P}'$.*

Proof Let $\mathcal{P}'$ be a refinement of $\mathcal{P}$ with $|\mathcal{P}'| < \epsilon$ defined by taking additional sections of I^n that have ν-measure zero. Apply the previous lemma to each rectangle $R \in \mathcal{P}$ taking for $R_1, \ldots, R_N$ the members of $\mathcal{P}'$ that are contained in R, with $\alpha_i = \nu(R_i)$. The homeomorphisms so obtained fit together to define $h \in \mathcal{H}[I^n, \partial I^n]$ with the required properties. $\square$

Proof of Homeomorphic Measures Theorem 9.1 As stated earlier we need only show that if μ is a nonatomic, locally positive, Borel probability measure which is zero on the boundary, then there is some homeomorphism $f \in \mathcal{H}[I^n, \partial I^n]$ such that $\mu = \lambda f$. For any homeomorphism $f \in \mathcal{H}[I^n, \partial I^n]$ and any $0 < \epsilon < 1$, let $\omega(f, \epsilon)$ be a positive number such that $\|f - fg\| < \epsilon$ whenever $g \in \mathcal{H}[I^n]$ and $\|g\| < \omega(f, \epsilon)$.

Since $\mu(I^n) = \lambda(I^n)$, we may apply Lemma A2.3 to the measure λ, and any rectangular subdivision $\mathcal{P}_1$ of I^n with $\mu(\partial R) = 0$ for each $R \in \mathcal{P}_1$, $|\mathcal{P}_1| < 1$, and the numbers $\alpha_i = \mu(R_i)$ for each $R_i \in \mathcal{P}_1$. The lemma gives us a homeomorphism $g_1 \in \mathcal{H}[I^n, \partial I^n]$ such that $\mu(R) = \lambda g_1(R)$ and $\mu(\partial R) = \lambda g_1(\partial R) = 0$ for each $R \in \mathcal{P}_1$.

Now using Lemma A2.4, there is a refinement $\mathcal{P}_2$ of $\mathcal{P}_1$ with $|\mathcal{P}_2| < \omega(g_1, 1/2)$ and an $h_1 \in \mathcal{H}[I^n, \partial I^n]$ such that $\mu h_1(R) = \lambda g_1(R)$ and $\mu h_1(\partial R) = \lambda g_1(\partial R) = 0$ for all $R \in \mathcal{P}_2$. Applying Lemma A2.4 alternately to triples of the form $\mu h_1 \cdots h_j$, $\lambda g_1 \cdots g_j$, $\mathcal{P}_{2j}$ and $\mu h_1 \cdots h_j$, $\lambda g_1 \cdots g_{j+1}$, $\mathcal{P}_{2j+1}$ we can inductively determine a sequence of rectangular subdivisions $\{\mathcal{P}_j\}$ of I^n, and sequences $\{g_j\}$ and $\{h_j\}$ in $\mathcal{H}[I^n, \partial I^n]$ such that for each $j \geq 1$

(i) $\mathcal{P}_{j+1}$ is a refinement of $\mathcal{P}_j$

(ii) $|\mathcal{P}_{2j}| < \omega(g_1 \cdots g_j, 1/2^j)$ and $|\mathcal{P}_{2j+1}| < \omega(h_1 \cdots h_j, 1/2^j)$

(iii) g_{j+1} leaves each member of $\mathcal{P}_{2j}$ invariant and h_{j+1} leaves each member of $\mathcal{P}_{2j+1}$ invariant

(iv) $\mu h_1 \cdots h_j(R) = \lambda g_1 \cdots g_j(R)$ and $\mu h_1 \cdots h_j(\partial R) = \lambda g_1 \cdots g_j(\partial R)$ $= 0$ for $R \in \mathcal{P}_{2j}$.

It follows from (ii) and (iii) that the limits $g = \lim g_1 \cdots g_j$ and $h = \lim h_1 \cdots h_j$ exist in $\mathcal{H}[I^n, \partial I^n]$, and from (i), (iii) and (iv) that $\mu h(R) = \lambda g(R)$ for each $R \in \bigcup_1^\infty \mathcal{P}_{2j}$. Since the cells in $\bigcup_1^\infty \mathcal{P}_{2j}$ generate the Borel sigma algebra, it follows that $\mu h(U) = \lambda g(U)$ for every open set U and therefore for every Borel set in I^n. Thus $f = gh^{-1}$ fulfills the requirement of Theorem 9.1. $\square$

Finally we consider the Hilbert cube $I^\infty = \prod_{i=1}^\infty I_i$, where each I_i is the unit interval, and let λ denote product Lebesgue measure. Evidently if $\mu = \lambda h$ for a Borel measure μ on I^∞ and some homeomorphism h of I^∞, then μ must be a nonatomic, locally positive and normalized probability measure since λ has these properties.

What conditions on a Borel measure on I^∞ measure characterize when it is homeomorphic to Lebesgue measure? For the finite n-cube I^n, in order to characterize measures homeomorphic to λ it is necessary to require that the measure of the boundary is zero in addition to the conditions above. Since I^∞ has no boundary (it is topologically homogeneous) it is not surprising that the boundary condition can be dropped, but it is remarkable that no other condition is needed to replace it. This characterization was obtained by Oxtoby and Prasad [92] and independently by B. Weiss (unpublished) [106] and is given here as

Theorem A2.5 *A necessary and sufficient condition for a Borel measure μ on I^∞ to be homeomorphic to product Lebesgue measure on I^∞ is that*

(i) *μ is zero for points (i.e., nonatomic)*

(ii) *μ is a probability measure*

(iii) *μ is positive on open sets (μ is locally positive).*

The main problem is to consider what happens when a measure μ gives positive measure to the 'pseudo-boundary' $\delta I^\infty = I^\infty - \prod_{i=1}^\infty (0,1)$. The first step in the proof begins by constructing a homeomorphism h of I^∞ so that $\mu h_1(\delta I^\infty) = 0$. Then Oxtoby and Prasad use arguments similar to the finite dimensional case. The details can be found in [92]. The results of Part I for I^n, $2 \le n \le \infty$, apply equally for I^∞.

A2.3 Homeomorphic Measures on Compact Manifolds

Two immediate corollaries of the Homeomorphic Measures Theorem follow. The first uses the Brown map (see Theorem 9.6) to extend the Homeomorphic Measures Theorem from I^n to compact connected manifolds X.

Corollary A2.6 *Let X be a compact connected n-manifold ($n \geq 2$) and let ν_1 and ν_2 be OU measures on X with $\nu_1(X) = \nu_2(X)$. Then there is a homeomorphism h of X such that $\nu_1(U) = \nu_2 h(U)$ for all Borel subsets U of X. Furthermore the homeomorphism h can always be chosen to fix any set of singular points of X obtained from boundary identifications of I^n. In particular, because the boundary of the manifold is always a subset of any set of singular points, h can always be chosen to fix the boundary of X.*

Proof Consider the following probability measure on X: for each Borel set $U \subset X$, define

$$\mu(U) = \frac{1}{2\nu_1(X)}(\nu_1(U) + \nu_2(U)).$$

Let ϕ be a Brown map from (I^n, λ) to (X, μ) from Theorem 9.6 such that $\lambda = \mu\phi$ and $\mu\phi(\partial I^n) = 0$. Then in particular, $\nu_1\phi(\partial I^n) = \nu_2\phi(\partial I^n) = 0$. Consequently the two measures $\mu_i = \nu_i\phi$, $i = 1, 2$, are both nonatomic, locally positive Borel measures on I^n which vanish on the boundary of I^n, and satisfy $\mu_1(I^n) = \mu_2(I^n) = c\lambda(I^n)$, where the constant $c = \mu_1(I^n)$. Thus the Homeomorphic Measures Theorem (Theorem 9.1) for I^n implies that there are homeomorphisms $h_1, h_2 \in \mathcal{H}[I^n, \partial I^n]$, such that $\mu_1 = c\lambda h_1$ and $\mu_2 = c\lambda h_2$. It is now a simple exercise to verify that the homeomorphism h of X defined by $h = \phi h_2^{-1} h_1 \phi^{-1}$ satisfies the requirements of the corollary. $\square$

The second corollary to the Homeomorphic Measures Theorem answers the following question: given two Borel probability measures, one on the square and the other on the disk, is it possible to find a measure preserving homeomorphism between these two spaces? The answer is yes if both of these measures are OU probability measures. More generally

Corollary A2.7 *Let X_1 and X_2 be compact connected n-manifolds and let g be any homeomorphism from X_1 to X_2. Let ν_i, $i = 1, 2$, be OU measures on X_i such that $\nu_1(X_1) = \nu_2(X_2)$. Then there is a homeomorphism*

f of X_1 to X_2 which carries ν_1 to ν_2 and which is equal to g for all boundary points of X_1.

Proof The measure $\mu_1(U) = \nu_2 g(U)$ (for Borel sets $U \subset X_1$) is a nonatomic locally positive Borel measure, zero on the boundary of X_1 and such that $\mu_1(X_1) = \nu_1(X_1)$. Hence by the previous corollary, there is a homeomorphism h of X_1 such that $\nu_1 = \mu_1 h$, which leaves fixed all singular points and hence all boundary points of X_1. Then $f = gh$ is a homeomorphism from X_1 to X_2 that agrees with g on boundary points of X_1 and satisfies $\nu_1 = \mu_1 h = \nu_2 f$. $\square$

Observe that we can sharpen the corollary above in the following manner. If Y_i is a locally flat codimension 0 submanifold of Int X_i for $i = 1, 2$ and g restricts to a homeomorphism of $Y_1 \to Y_2$ which is already measure preserving (i.e., $\nu_1|_{Y_1} = \nu_2 g|_{Y_2}$), then f can be chosen to equal g on Y_1. Just apply the corollary to $g : X_1 - \text{Int } Y_1 \to X_2 - \text{Int } Y_2$.

An extension of Brown's Theorem (Theorem 9.3) to Hilbert cube manifolds in [97] by the second author, gives analogous theorems for homeomorphic measures on Hilbert cube manifolds.

A2.4 Homeomorphic Measures on Noncompact Manifolds

At the end of Chapter 14 (Noncompact Manifolds and Ends) we noted that the Homeomorphic Measures Theorem for noncompact manifolds requires some condition on the behavior of the measures at the ends. In this section we outline R. Berlanga and D. Epstein's proof [38] of their result that two OU measures, μ and ν, on a noncompact connected n-manifold X, are equivalent by an end preserving homeomorphism if $\mu(X) = \nu(X)$ and they are infinite on the same set of ends. We use the notions associated with ends given in Chapter 14.

Theorem A2.8 (Berlanga–Epstein [38]) *Let X be a sigma compact, connected n-manifold ($n \geq 2$) and let μ and ν be two OU measures on X (i.e., nonatomic, Borel measures, positive on open sets, which are zero on the boundary of X). Suppose $\mu(X) = \nu(X)$ and furthermore suppose that the measures are infinite on the same set of ends (i.e., the measures μ^* and ν^*, induced on the ends $E[X]$, are identical). Then there is a homeomorphism h of X such that $\nu = \mu h$. Furthermore, h can be chosen to fix the boundary of X (i.e., $h \in \mathcal{H}[X, \partial X]$), and be end preserving.*

Actually, Berlanga and Epstein also observe that the homeomorphism h can be taken to be isotopic to the identity.

A simple outline of their proof is given by the following. Suppose that we have a relative n-cell K_1 such that $\mu(K_1) = \nu(K_1)$, and that μ and ν restricted to K_1 are OU measures on K_1. Then the compact version of the Homeomorphic Measures Theorem (Corollary A2.6) can be applied on K_1, to find a homeomorphism h_1 of K_1 which 'corrects the measure μ on K_1' (i.e., $\mu h_1 = \nu$ when restricted to Borel subsets of K_1). Furthermore if we can choose K_1 so that $\mu(e(K_1)) = \nu(e(K_1))$ for all ends $e \in E[X]$, then when we consider the measures μ and ν restricted to each component $e(K_1)$, we will be in exactly the same situation as before: i.e., each $e(K_1)$ is a noncompact manifold with two OU measures having the same total measure, and infinite on the same set of ends. The idea then will be to select a larger relative n-cell K_2 containing K_1 in its interior so that the μ and ν measures of K_2 and of each component $e(K_2)$ are the same. The Homeomorphic Measures Theorem will then be applied on the relative n-cell K_2, to find a homeomorphism h_2 of K_2 so that μh_2 and ν are identical measures on K_2. We will find that we can choose h_2 so that its support lies outside of K_1. In this manner, we will define our required homeomorphism taking μ to ν successively on larger compact subsets of X. That we can choose relative n-cells with the properties described above is the content of the succeeding lemmas, the first set of which lays the topological foundations for the later measure theoretic results.

We conclude this section by noting that a smooth version of this theorem, due to R. Greene and K. Shiohama [67] (1979) preceded Berlanga and Epstein's theorem. Greene and Shiohama's theorem states that if X is a noncompact connected oriented manifold and if ω and τ are C^∞ volume forms on X with the same total volume and if the volume forms are infinite on the same set of ends then there is a diffeomorphism h of X such that $\omega h = \tau$. Greene and Shiohama's theorem is the noncompact analog of Moser's 1965 [87] smooth version for compact manifolds, of Oxtoby and Ulam's 1941 homeomorphic measures theorem (Corollary A2.6). Indeed we note that Berlanga and Epstein use Oxtoby and Ulam's Homeomorphic Measures Theorem (Corollary A2.6) in the same way that Greene and Shiohama use Moser's theorem for compact manifolds.

A2.5 Proof of the Berlanga–Epstein Theorem

For the reader's convenience we recall some of the notation from Chapter 14. For any $Y \subset X$ we define Y° to be the topological interior of Y, Int Y to be $(X - \partial X) \cap Y^\circ$, Cl Y to be the closure of Y in X, Fr Y to be the topological frontier of Y in X, and the boundary of Y, denoted Bdry Y, to be $(\partial X \cap Y) \cup$ Fr Y.

Recall that a set $K \subset X$ is called a *relative n-cell* if there exists a continuous function $\phi : I^n \to K$ such that

(i) ϕ is onto

(ii) ϕ restricted to Int I^n is a homeomorphism onto its image

(iii) $\phi^{-1}\phi(\partial I^n) = \partial I^n$.

Note that if μ_1 and μ_2 are two OU measures on a relative n-cell $K \subset X$, of the same total mass $(\mu_1(K) = \mu_2(K))$, then Corollary A2.6 implies that there is a homeomorphism $h \in \mathcal{H}[K, \text{Bdry } K]$ such that $\mu_1 h = \mu_2$.

The next two lemmas are purely topological and so we do not prove them – rather we refer the interested reader to Berlanga and Epstein's paper [38] for their proofs.

Lemma A2.9 *Let A, B be two disjoint compact sets in the sigma compact connected n-manifold X. Let μ be a sigma finite OU measure on X. Then there exists a finite disjoint family of relative n-cells $\{L_1, \ldots, L_r\}$ with the property that $B \subset \bigcup_{i=1}^r L_i^\circ$, $L_i \cap A = \emptyset$ for $i = 1, 2, \ldots, r$, and $\mu(\text{Bdry } L_i) = 0$ for $i = 1, 2, \ldots, r$.*

Furthermore, if $X - A$ is connected then the collection of relative n-cells can be chosen to contain only one element.

We end the topological preliminaries with some results on the complements of compact sets. To this end, let $K \subset X$ be a compact subset of X. Consider now the set of connected components of $X - K$. If $V \subset X - K$ is a connected component of $X - K$, then we say that V is *bounded* if its closure is compact and *unbounded* otherwise. Finally we define

$$\hat{K} = X - \bigcup \{V : V \text{ is an unbounded connected component of } X - K\}.$$

The following lemma is true in a more general setting than for a noncompact manifold but we need it only for the manifold setting.

Lemma A2.10 *Let $K \subset X$ be a compact subset of X. Then $X - K$ has* *only finitely many unbounded components and $\hat{K}$ is a compact set whose complement has only unbounded components.*

Finally, recall that each compact set K determines an equivalence relation $\sim_K$ on the ends $E[X]$, and that $\mathcal{P}_K$ is the finite partition of $E[X]$ determined by the unbounded connected components of $X - K$. Furthermore when $K = \hat{K}$, then the connected components of $X - K$ are just $\{P(K) : P \in \mathcal{P}_K\}$.

The next lemma is basic to the main construction of the theorem in that it allows us to change the measure of any finite number of sets of finite positive measure by some homeomorphism of X with compact support so that the given sets have prescribed (nonzero) measures.

Lemma A2.11 *Let X be a connected manifold and let μ be a sigma finite OU measure on X. Let $\beta_1, \beta_2, \ldots, \beta_k$ be positive numbers and let $V_1, V_2, \ldots, V_k$ be a disjoint family of Borel sets in X such that*

(1) $0 < \mu(V_i) < \infty$ for each $i = 1, 2, \ldots, k$

and either

(2) $0 < \mu(X - \bigcup_{i=1}^{k} V_i)$ and $\sum_{i=1}^{k} \beta_i < \mu(X)$

or

(3) $\mu(X - \bigcup_{i=1}^{k} V_i) = 0$ and $\sum_{i=1}^{k} \beta_i = \mu(X)$ (and hence $\mu(X) < \infty$).

Then there exists a compactly supported $h \in \mathcal{H}[X, \partial X]$ such that for each $i = 1, \ldots, k$, $\mu(h(V_i)) = \beta_i$, and μ and μh have the same sets of measure zero.

Proof By omitting one of the sets V_i, case (3) is reduced to case (2) and so we restrict ourselves to this case.

Set $V_0 = X - \bigcup_{i=1}^{k} V_i$, so that $\{V_0, V_1, \ldots, V_k\}$ forms a partition of X. By the inner regularity of the measure μ, we can find for each $i = 0, \ldots, k$, compact sets $F_i \subset V_i$ such that

(i) $\mu(F_i) > 0$ for $i = 0, \ldots, k$

(ii) $\mu(V_i - F_i) < \beta_i$ for $i = 1, \ldots, k$.

Let F_{k+1} be any compact set with $\mu(F_{k+1}) > \sum_{i=1}^{k} \beta_i$. Apply Lemma A2.9 to $B = \bigcup_{i=0}^{k+1} F_i$ and $A = \emptyset$ (so that $X - A$ is connected) to obtain a relative n-cell K with $\mu(\text{Bdry } K) = 0$ such that B is contained in K°, and K satisfies the following three properties:

(i) $\mu(V_i \cap K) > 0$ for $i = 0, \ldots, k$

(ii) $\mu(K) > \sum_{i=1}^{k} \beta_i$ and

(iii) $\mu(V_i - K) < \beta_i$ for $i = 1, \ldots, k$.

Let

$\alpha_i = \beta_i - \mu(V_i - K)$ for each $i = 1, \ldots, k$, and

$\alpha_0 = \mu(K) - \sum_{i=1}^{k} \alpha_i$.

Note that $\alpha_i > 0$ for each $i = 0, \ldots, k$, and $\sum_{i=0}^{k} \alpha_i = \mu(K)$. For each $i = 0, 1, \ldots, k$, define a probability measure μ_i on K to be the conditional probability of μ conditioned on V_i; i.e., each μ_i is the probability measure on K defined by setting, for each Borel set $U \subset K$,

$$\mu_i(U) = \frac{\mu(U \cap K \cap V_i)}{\mu(K \cap V_i)}.$$

By taking a weighted sum of these conditional measures we define a new OU measure ν with support in K, defined by the formula

$$\nu(U) = \sum_{i=0}^{k} \alpha_i \mu_i(U)$$

for all Borel sets $U \subset K$. We note that ν is indeed an OU measure on the relative n-cell K: first, ν is positive on open subsets of K, because $\{K \cap V_i : i = 0, \ldots, k\}$ forms a partition of K; since each μ_i is nonatomic and zero on Bdry K, then so too is ν – indeed, the measure ν has the same sets of zero measure on K as μ does; and, finally we note that $\nu(V_i \cap K) = \alpha_i$ for $i = 0, 1, \ldots, k$. Thus since $\nu(K) = \sum_{i=0}^{k} \alpha_i = \mu(K)$, ν and μ are two OU measures on the relative n-cell K of the same total finite measure. So by the compact version (Corollary A2.6) of the Homeomorphic Measures Theorem, we can find $h \in \mathcal{H}[K, \text{Bdry } K]$, such that $\mu h|_K = \nu|_K$ (i.e., the restrictions of these measures to K). Extend h to a homeomorphism of all of X by making it the identity off K. Note that for $i = 1, \ldots, k$,

$$
\begin{aligned}
\mu(h(V_i)) &= \mu(h(V_i) \cap K) + \mu(h(V_i) - K) \\
&= \nu(V_i \cap K) + \mu(V_i - K) \\
&= \alpha_i + (\beta_i - \alpha_i) = \beta_i
\end{aligned}
$$

as required.   □

The Homeomorphic Measures Theorem for the noncompact manifold X will be proved by repeatedly applying the next lemma. The lemma shows that whenever we have two OU measures μ_1 and μ_2 on X which are identical as OU measures restricted to some compact set K, then we can find a larger compact set L and a homeomorphism h of X such that the measures $\mu_1 h$ and μ_2 when restricted to (Borel subsets of) L are identical. The exact statement is:

Lemma A2.12 *Let X be a sigma compact, connected manifold and let A, B, and K be compact subsets of X such that $A \subset K^\circ$ and the complement of K has only unbounded components (i.e., $K = \hat{K}$). Let μ_1, μ_2 be OU measures on X which agree on ends and suppose further that they are identical as measures on K $(\mu_1|_K = \mu_2|_K)$, and $\mu_1(P(K)) = \mu_2(P(K))$ for every $P \in \mathcal{P}_K$. Then there exists an $h \in \mathcal{H}[X, \partial X]$ with compact support and a compact set L such that*

  (i) supp $h \cap A = \emptyset$

  (ii) $L^\circ \supset K \cup B$ and $L = \hat{L}$

  (iii) $\mu_1 h|_L = \mu_2|_L$

  (iv) $\mu_1 h_1(P(L)) = \mu_2(P(L))$ for $P \in \mathcal{P}_L$, i.e., for all connected components $P(L)$ of $X - L$.

Proof Let C be a compact set in X such that $K \cup B \subset C^\circ$. Applying Lemma A2.9 to the compact sets A and $\hat{C} - K^\circ$ with $\mu = \mu_1 + \mu_2$, we find a disjoint family of relative n-cells $L_1, \ldots, L_r$ such that

  (i) $\mu_1(\text{Bdry } L_i) = \mu_2(\text{Bdry } L_i) = 0$ for each $i = 1, 2, \ldots, r$

  (ii) $L_i \cap A = \emptyset$ for each $i = 1, 2, \ldots, r$

  (iii) $\hat{C} \subset K^\circ \bigcup_{i=1}^{r} L_i^\circ$.

Let $P(K)$ be any connected (unbounded) component of $X - K$ and let $W_1, \ldots, W_q$ be the components of $P(K) - \hat{C}$. Observe that because the measures agree on ends, $\mu_1(W_j) = \infty$ if and only if $\mu_2(W_j) = \infty$. The connected component $P(K)$ of $X - K$ is partitioned into the following collection of sets

$L_i \cap P(K) - \bigcup_{j=1}^{q} W_j$; $i = 1, 2, \ldots, r$
$L_i \cap W_j$; $i = 1, 2, \ldots, r$; $j = 1, 2, \ldots, q$
$W_j - \bigcup_{i=1}^{r} L_i$; $j = 1, 2, \ldots, q$

Lemma A2.11 says that there is a homeomorphism h_1 with compact support in $P(K)$ that changes the μ_1 measure of the sets above. Thus we may assume without loss of generality that the $\mu_1 h_1$ and μ_2 measure of each of the sets in the decomposition above have the same total measure. Clearly we can do this simultaneously for each connected component $P(K)$ of $X - K$ and denote the resulting homeomorphism of X so obtained, again by h_1.

    Thus we can assume that $\mu_1 h_1(L_i) = \mu_2(L_i)$ for each $i = 1, 2, \ldots, r$. Since each L_i is a relative n-cell (hence, the Homeomorphic Measures Theorem is true for OU measures $\mu_1 h_1$ and μ_2 on each L_i), there is a homeomorphism h_2 of X that $\mu_1 h_1 h_2$ and μ_2 are identical measures on $\bigcup_{i=1}^{r} L_i$ (which contains the support of $h_1 h_2$), hence in all of $K \cup \bigcup_{i=1}^{r} L_i$.

Since the latter set contains $\hat{C}$, setting $L = \hat{C}$ and $h = h_1 h_2$, we have $\mu_1 h|_L = \mu_2|_L$ and $\mu_1 h(W) = \mu_2(W)$ for all connected components W of $X - L$. This concludes the proof. $\qquad\square$

Now we are in a position to prove Berlanga and Epstein's Theorem:

Proof of Theorem A2.8 Write $X = \bigcup_{i=1}^{\infty} C_i$, where C_i is compact and $C_i \subset C_{i+1}^{\circ}$ for each i.

Step 1: Using the previous lemma with $A_0 = \emptyset$, $B_0 = C_1$, $K_0 = \emptyset$, $\mu_1 = \mu$ and $\mu_2 = \nu$, we obtain a compactly supported homeomorphism $h_1 \in \mathcal{H}[X, \partial X]$ and a compact set L_1 such that

(i) $L_1^{\circ} \supset K_0 \cup B_0 = C_1$ and $L_1 = \hat{L}_1$
(ii) $\mu h_1|_{L_1} = \nu|_{L_1}$
(iii) $\mu h_1(P(L_1)) = \nu(P(L_1))$ for all $P \in \mathcal{P}_{L_1}$.

Step 2: Letting $A_1 = C_1$, $B_1 = C_2 \cup \operatorname{supp} h_1$, $K_1 = L_1$, $\mu_1 = \nu$ and $\mu_2 = \mu h_1$, and again using Lemma A2.12, we obtain an $h_2 \in \mathcal{H}[X, \partial X]$ with compact support and a compact set L_2 such that

(i) $\operatorname{supp} h_2 \cap C_1 = \emptyset$
(ii) $L_2^{\circ} \supset L_1 \cup C_2 \cup \operatorname{supp} h_1$ and $L_2 = \hat{L}_2$
(iii) $\mu h_1|_{L_2} = \nu h_2|_{L_2}$
(iv) $\mu h_1(P(L_2)) = \nu h_2(P(L_2))$ for all connected components $P(L_2)$ of $X - L_2$.

Step 3: Using Lemma A2.12 with $A_2 = L_1 \cup C_2 \cup \operatorname{supp} h_1$, $B_2 = C_3 \cup \operatorname{supp} h_2$, $K_2 = L_2$, $\mu_1 = \mu h_1$, $\mu_2 = \nu h_2$ we get a compact set L_3 and a compactly supported homeomorphism h_3 such that

(i) $\operatorname{supp} h_3 \cap (L_1 \cup C_2 \cup \operatorname{supp} h_1) = \emptyset$
(ii) $L_3^{\circ} \supset L_2 \cup C_3 \cup \operatorname{supp} h_2$ and $L_3 = \hat{L}_3$
(iii) $\mu h_1 h_3|_{L_3} = \nu h_2|_{L_3}$
(iv) $\mu h_1 h_3(P(L_3)) = \nu h_2(P(L_3))$ for all connected components $P(L_3)$ of $X - L_3$.

Note that the homeomorphisms h_1 and h_3 have disjoint supports.

Step $(i + 1)$: Continuing in this manner suppose we have inductively defined a sequence of compact sets $L_0 = \emptyset, L_1, L_2, \ldots, L_i$, with $L_i = \hat{L}_i$, and compactly supported homeomorphisms $h_0 = id, h_1, h_2, \ldots, h_i$. Applying Lemma A2.12 to $A_i = L_{i-1} \cup C_i \cup \operatorname{supp} h_{i-1}$, $B_i = C_{i+1} \cup \operatorname{supp} h_i$ and $K_i = L_i$ we get a compactly supported homeomorphism h_{i+1} and a compact subset L_{i+1} with $L_{i+1} = \hat{L}_{i+1}$ such that

(i) $\operatorname{supp} h_{i+1} \cap A_i = \operatorname{supp} h_{i+1} \cap (L_{i-1} \cup C_i \cup \operatorname{supp} h_{i-1}) = \emptyset$

(ii) $(L_{i+1})^\circ \supset A_i$

(iii) When i is even the measures on L_{i+1},

$$\mu(h_{i+1} \cdot h_{i-1} \cdots \cdots h_1)|_{L_{i+1}} = \nu(h_i \cdot h_{i-2} \cdots \cdots h_0)|_{L_{i+1}},$$

and when i is odd,

$$\mu(h_i \cdot h_{i-2} \cdots \cdots h_1)|_{L_{i+1}} = \nu(h_{i+1} \cdot h_{i-1} \cdots \cdots h_0)|_{L_{i+1}}.$$

(iv) For $P \in \mathcal{P}_{L_{i+1}}$, the measures of the components $P(L_{i+1})$ satisfy

$$\mu(h_{i+1} \cdot h_{i-1} \cdots \cdots h_1)(P(L_{i+1})) = \nu(h_i \cdot h_{i-2} \cdots \cdots h_0)(P(L_{i+1}))$$

when i is even, and when i is odd

$$\nu(h_{i+1} \cdot h_{i-1} \cdots \cdots h_0)(P(L_{i+1})) = \mu(h_i \cdot h_{i-2} \cdots \cdots h_1)(P(L_{i+1})).$$

Because the even indexed homeomorphisms have disjoint supports there is no problem in defining the infinite composition

$$h_e = \lim_{i \to \infty} h_{2i} \cdot h_{2i-2} \cdots \cdots h_0.$$

Similarly, the composition of the odd indexed homeomorphisms is well defined

$$h_o = \lim_{i \to \infty} h_{2i+1} \cdot h_{2i-1} \cdots \cdots h_1.$$

And so $h_o, h_e \in \mathcal{H}[X, \partial X]$ and $\mu h_o = \nu h_e$. Setting $h = h_o h_e^{-1}$ we get the required homeomorphism such that $\mu h = \nu$. $\qquad \square$

Finally we note that in [39], R. Berlanga has proved a generalization, to sigma compact connected manifolds X, of M. Brown's Theorem 9.3 for compact manifolds. In this setting (when X is a sigma compact connected n-manifold ($n \geq 2$)), we have seen that the set of ends $E[X]$ is a totally disconnected compact metrizable space; thus we can construct some subset E in the boundary of the n-cube, such that $I^n - E$ and X are two n-manifolds with the same set of ends. Berlanga's Theorem shows that X can be viewed as the identification space of $I^n - E$ obtained by identifying points within $\partial I^n - E$ to other points in $\partial I^n - E$. More precisely Berlanga's generalization of Brown's Theorem is

Theorem A2.13 *Let X be a sigma compact connected n-manifold. Then there is a compact set $E \subset \partial I^n$ and a continuous map $\phi : I^n - E \to X$ such that*

(i) *ϕ is a continuous proper map onto X (i.e., ϕ is proper means that the inverse image of a compact set in X is compact in $I^n - E$)*

(ii) $\phi|\operatorname{Int} I^n$ is a homeomorphism of the interior of I^n onto its image.

(iii) $\phi(\partial I^n - E) \cap \phi(\operatorname{Int} I^n) = \emptyset$ and $\phi(\partial I^n - E)$ has empty interior.

(iv) ϕ extends naturally to $\tilde{\phi} : I^n \to X \cup E[X]$ in such a way that the restriction of $\tilde{\phi}$ to E is a homeomorphism from E onto $E[X]$.

Furthermore, E can be chosen to be contained in the 'top of the nth coordinate face', i.e., a subset of $[1/3, 2/3] \times (1/2, 1/2, \ldots, 1/2, 1)$. In particular if $E[X]$ has no isolated points then E can be chosen to be a 'standard' copy of the Cantor ternary set in ∂I^n.

Berlanga notes that when the dimension n is 1 or 2, this theorem follows from the classification of second countable manifolds (see Ahlfors and Sario [3]). For dimensions $n \geq 3$, Berlanga's argument, when $E[X] = \emptyset$ (so that X is compact), reduces to those given in Brown's proof of Theorem 9.3 (in [45]).

Berlanga's Structure Theorem can be used to study measure preserving homeomorphisms of sigma compact manifolds in the same manner that we used Brown's Theorem to study measure preserving homeomorphisms of compact manifolds (see Chapters 9 and 10). Thus for example, OU measures on the manifold X give rise OU measures on $I^n - E$.

Bibliography

[1] Jon Aaronson. *An introduction to infinite ergodic theory*, volume 50 of *Mathematical Surveys and Monographs*. American Mathematical Society, Providence, RI, 1997.

[2] J. Aarts and Fons Daalderop. Chaotic homeomorphisms on manifolds. *Topology and its Applications*, 96:93–96, 1999.

[3] Lars V. Ahlfors and Leo Sario. *Riemann surfaces*. Princeton University Press, Princeton, NJ, 1960. Princeton Mathematical Series, No. 26.

[4] Ethan Akin. Stretching the Oxtoby–Ulam theorem. *Colloquium Math. (Iwanik volume)*, 84:83–94, 2000.

[5] Steve Alpern. *On Approximating Measure Preserving Homeomorphisms*. PhD thesis, Courant Institute of Mathematical Sciences, New York University, New York, 1973.

[6] Steve Alpern. A new proof of the Oxtoby–Ulam theorem. In J. Moser, E. Phillips, and S. Varadhan, editors, *Ergodic theory*, pages 125–131. Courant Institute of Mathematical Sciences New York University, New York, 1975. MR58:6177.

[7] Steve Alpern. New proofs that weak mixing is generic. *Invent. Math.*, 32(3):263–278, 1976. MR53:5828.

[8] Steve Alpern. Approximation to and by measure preserving homeomorphisms. *J. London Math. Soc. (2)*, 18(2):305–315, 1978. MR80d:28033.

[9] Steve Alpern. Superhamiltonian graphs. *J. Combin. Theory Ser. B*, 25(1):62–73, 1978. MR58:21825.

[10] Steve Alpern. A topological analog of Halmos' conjugacy lemma. *Invent. Math.*, 48(1):1–6, 1978. MR80d:28034.

[11] Steve Alpern. Generic properties of measure preserving homeomorphisms. In *Ergodic theory (Proc. Conf., Math. Forschungsinst., Oberwolfach, 1978)*, volume 729 of *Lecture Notes in Math.*, pages 16–27. Springer, Berlin, 1979. MR80m:28019.

[12] Steve Alpern. Measure preserving homeomorphisms of R^n. *Indiana Univ. Math. J.*, 28(6):957–960, 1979. MR80m:28014b.

[13] Steve Alpern. Return times and conjugates of an antiperiodic transformation. *Ergodic Theory Dynamical Systems*, 1(2):135–143, 1981. MR83i:28020.

[14] Steve Alpern. Nonstable ergodic homeomorphisms of R^4. *Indiana Univ. Math. J.*, 32(2):187–191, 1983. MR84d:28027.

[15] Steve Alpern. Area-preserving homeomorphisms of the open disk without fixed points. *Proc. Amer. Math. Soc.*, 103(2):624–626, 1988. MR89e:55003.

[16] Steve Alpern. Almost periodic ergodic R^n-homeomorphisms. *Adv. Math.*, 116(1):46–54, 1995. MR97e:28010.

[17] Steve Alpern. Combinatorial approximation by Devaney-chaotic or periodic volume preserving homeomorphisms. *International Journal of Bifurcation and Chaos*, 9(5):843–848, 1999.

[18] Steve Alpern, J. R. Choksi, and V. S. Prasad. Conjugates of infinite measure preserving transformations. *Canad. J. Math.*, 40(3):742–749, 1988. MR89m:28026.

[19] Steve Alpern and Robert D. Edwards. Lusin's theorem for measure-preserving homeomorphisms. *Mathematika*, 26(1):33–43, 1979. MR81j:28025.

[20] Steve Alpern and V. S. Prasad. End behaviour and ergodicity for homeomorphisms of manifolds with finitely many ends. *Canad. J. Math.*, 39(2):473–491, 1987. MR89a:28013.

[21] Steve Alpern and V. S. Prasad. Weak mixing manifold homeomorphisms preserving an infinite measure. *Canad. J. Math.*, 39(6):1475–1488, 1987. MR89a:28014.

[22] Steve Alpern and V. S. Prasad. Dynamics induced on the ends of a noncompact manifold. *Ergodic Theory Dynamical Systems*, 8(1):1–15, 1988. MR89e:58069.

[23] Steve Alpern and V. S. Prasad. Coding a stationary process to one with prescribed marginals. *Ann. Probab.*, 17(4):1658–1663, 1989. MR91f:28007.

[24] Steve Alpern and V. S. Prasad. Return times for nonsingular measurable transformations. *J. Math. Anal. Appl.*, 152(2):470–487, 1990. MR91m:28023.

[25] Steve Alpern and V. S. Prasad. Typical recurrence for lifts of mean rotation zero annulus homeomorphisms. *Bull. London Math. Soc.*, 23(5):477–481, 1991. MR93a:28012.

[26] Steve Alpern and V. S. Prasad. Combinatorial proofs of the Conley–Zehnder–Franks theorem on a fixed point for torus homeomorphisms. *Adv. Math.*, 99(2):238–247, 1993. MR94c:58101.

[27] Steve Alpern and V. S. Prasad. Topological ergodic theory and mean rotation. *Proc. Amer. Math. Soc.*, 118(1):279–284, 1993. MR93f:58130.

[28] Steve Alpern and V. S. Prasad. Typical transitivity for lifts of rotationless annulus or torus homeomorphisms. *Bull. London Math. Soc.*, 27(1):79–81, 1995. MR96g:58097.

[29] Steve Alpern and V. S. Prasad. Chaotic homeomorphisms of R^n lifted from torus homeomorphisms. *Bull. London Math. Soc.*, 31(5):577–580, 1999.

[30] Steve Alpern and V. S. Prasad. Maximally chaotic homeomorphisms of sigma-compact manifolds. *Topology and its Applications*, 105:103–112, 2000.

[31] Steve Alpern and V. S. Prasad. Extensions of the Oxtoby–Ulam theorem on the prevalence of ergodicity for measure preserving

homeomorphisms, *submitted.* Available in LSE preprint series
LSE-CDAM-2000-06.

[32] Stephen A. Andrea. On homeomorphisms of the plane which have no
fixed points. *Abh. Math. Sem. Univ. Hamburg,* 30:61–74, 1967.
MR34:8397.

[33] L. Antoine. Sur l'homéomorphie de deux figures et de leurs voisinages.
J. Math. Pures Appli., 4:221–325, 1921.

[34] V. I. Arnold. Fixed points of symplectic diffeomorphisms. *Proc.
Sympos. Pure Math.,* 28:6, 1976.

[35] V. I. Arnold. *Mathematical methods of classical mechanics,* volume 60
of *Graduate Texts in Mathematics.* Springer-Verlag, New York, 1989.
Translated from the 1974 Russian original by K. Vogtmann and A.
Weinstein. Corrected reprint of the second (1989) edition,
MR96c:70001.

[36] Daniel Asimov. On volume-preserving homeomorphisms of the open
n-disk. *Houston J. Math.,* 2(1):1–3, 1976. MR53:4147.

[37] J. Banks, J. Brooks, G. Cairns, G. Davis, and P. Stacey. On Devaney's
definition of chaos. *Amer. Math. Monthly,* 99(4):332–334, 1992.

[38] R. Berlanga and D. B. A. Epstein. Measures on sigma-compact
manifolds and their equivalence under homeomorphisms. *J. London
Math. Soc. (2),* 27(1):63–74, 1983. MR84m:28023.

[39] Ricardo Berlanga. A mapping theorem for topological sigma-compact
manifolds. *Compositio Math.,* 63(2):209–216, 1987. MR89d:57021.

[40] A. S. Besicovitch. A problem on topological transformation of the
plane. *Fund. Math.,* 28:61–65, 1937.

[41] G. D. Birkhoff. Proof of Poincaré's last geometric theorem. *Trans.
Amer. Math. Soc.,* 14:14–22, 1913.

[42] Garrett Birkhoff. Three observations on linear algebra. *Univ. Nac.
Tucumán. Revista A.,* 5:147–151, 1946. MR 8,561a.

[43] D. G. Bourgin. Homeomorphisms of the open disk. *Studia Math.,*
31:433–438, 1968. MR38:3852.

[44] L. E. J. Brouwer. Beweis des ebenen Translationssatzes. *Math. Ann.,*
72:37–54, 1912.

[45] Morton Brown. A mapping theorem for untriangulated manifolds. In
*Topology of 3-manifolds and related topics (Proc. The Univ. of Georgia
Institute, 1961),* pages 92–94. Prentice-Hall, Englewood Cliffs, NJ,
1962. MR28:1599.

[46] Morton Brown and Herman Gluck. Stable structures on manifolds i.
Ann. of Math., 79:1–17, 1964.

[47] G. Cairns, G. Davis, D. Elton, A. Kolganova, and P. Perversi. Chaotic
group actions. *Enseign. Math. (2),* 41(1–2):123–133, 1995.

[48] Grant Cairns, Barry Jessup, and Marcel Nicolau. Topologically
transitive homeomorphisms of quotients of tori. *Discrete Contin.
Dynam. Systems,* 5(2):291–300, 1999.

[49] Grant Cairns and Alla Kolganova. Chaotic actions of free groups.
Nonlinearity, 9(4):1015–1021, 1996.

[50] J. R. Choksi and Shizuo Kakutani. Residuality of ergodic measurable
transformations and of ergodic transformations which preserve an
infinite measure. *Indiana Univ. Math. J.,* 28(3):453–469, 1979.
MR80d:28042.

[51] J. R. Choksi and V. S. Prasad. Approximation and Baire category theorems in ergodic theory. In *Measure theory and its applications (Sherbrooke, Que., 1982)*, volume 1033 of *Lecture Notes in Math.*, pages 94–113. Springer, Berlin, 1983. MR85d:28011.

[52] Charles O. Christenson and William L. Voxman. *Aspects of topology.* Marcel Dekker Inc., New York, 1977. Pure and applied Mathematics, Vol. 39.

[53] C. C. Conley and E. Zehnder. The Birkhoff–Lewis fixed point theorem and a conjecture of V. I. Arnold. *Invent. Math.*, 73(1):33–49, 1983. MR85e:58044.

[54] I. P. Cornfeld, S. V. Fomin, and Ya. G. Sinaĭ. *Ergodic theory*, volume 245 of *Grundlehren der Mathematischen Wissenschaften [Fundamental Principles of Mathematical Science]*. Springer-Verlag, New York, 1982. Translated from the Russian by A. B. Sosinskiĭ.

[55] A. Daalderop and R. Fokkink. Chaotic homeomorphisms are generic. *Top. and Appl.*, 102:297–302, 2000.

[56] Robert L. Devaney. *A first course in chaotic dynamical systems.* Addison-Wesley Publishing Company Advanced Book Program, Reading, MA, 1992. MR94a:58124.

[57] S. J. Eigen and V. S. Prasad. Multiple Rokhlin tower theorem: a simple proof. *New York J. Math.*, 3A (Proceedings of the New York Journal of Mathematics Conference, June 9–13, 1997):11–14 (electronic), 1997/98. e-journal article at http://nyjm.albany.edu:8000/j/1997/3A_11.html.

[58] A. Fathi. Structure of the group of homeomorphisms preserving a good measure on a compact manifold. *Ann. Sci. Ecole Norm. Sup. (4)*, 13(1):45–93, 1980. MR81k:58042.

[59] Albert Fathi. An orbit closing proof of Brouwer's lemma on translation arcs. *Enseign. Math. (2)*, 33(3–4):315–322, 1987. MR89d:55004.

[60] Martin Flucher. Fixed points of measure preserving torus homeomorphisms. *Manuscripta Math.*, 68(3):271–293, 1990. MR91j:58129.

[61] John Franks. Generalizations of the Poincaré–Birkhoff theorem. *Ann. of Math. (2)*, 128(1):139–151, 1988. MR89m:54052.

[62] John Franks. Recurrence and fixed points of surface homeomorphisms. *Ergodic Theory Dynamical Systems*, 8 (Charles Conley Memorial Issue):99–107, 1988. MR90d:58124.

[63] Nathaniel A. Friedman. *Introduction to ergodic theory.* Van Nostrand Reinhold Co., New York, 1970. Van Nostrand Reinhold Mathematical Studies, No. 29.

[64] H. Furstenberg. *Recurrence in ergodic theory and combinatorial number theory.* Princeton University Press, Princeton, NJ, 1981. M. B. Porter Lectures.

[65] C. Goffman, T. Nishiura, and D. Waterman. *Homeomorphisms in Analysis*, volume 54 of *Mathematical Surveys and Monographs.* American Mathematical Society, Providence, RI, 1997.

[66] Casper Goffman and George Pedrick. A proof of the homeomorphism of Lebesgue–Stieltjes measure with Lebesgue measure. *Proc. Amer. Math. Soc.*, 52:196–198, 1975. MR51:13170.

[67] R. E. Greene and K. Shiohama. Diffeomorphisms and volume-preserving embeddings of noncompact manifolds. *Trans. Amer. Math. Soc.*, 255:403–414, 1979. MR80k:58031.

[68] Marshall Hall Jr. *Combinatorial theory*. Wiley-Interscience Series in Discrete Mathematics. John Wiley & Sons Inc., New York, second edition, 1986. A Wiley-Interscience Publication; See Chapter 7 for P. Hall's theorem. MR87j:05001.

[69] Paul R. Halmos. Approximation theories for measure preserving transformations. *Trans. Amer. Math. Soc.*, 55:1–18, 1944. MR5:189b.

[70] Paul R. Halmos. In general a measure preserving transformation is mixing. *Ann. of Math. (2)*, 45:786–792, 1944. MR6:131d.

[71] Paul R. Halmos. *Measure Theory*. D. Van Nostrand Company, Inc., New York, 1950. MR11:504d.

[72] Paul R. Halmos. *Lectures on ergodic theory*. The Mathematical Society of Japan, 1956. Publications of the Mathematical Society of Japan, no. 3.

[73] John G. Hocking and Gail S. Young. *Topology*. Dover Publications Inc., New York, second edition, 1988.

[74] Paul G. Hoel, Sidney C. Port, and Charles J. Stone. *Introduction to stochastic processes*. Waveland Press Inc., 1987. Reprint of the 1972 book published by Houghton Mifflin, Boston.

[75] Shizuo Kakutani. Induced measure preserving transformations. *Proc. Imp. Acad. Tokyo*, 19:635–641, 1943. MR7:255f.

[76] A. B. Katok and A. M. Stepin. Metric properties of homeomorphisms that preserve measure. *Uspehi Mat. Nauk*, 25:193–220, 1970. English translation in Russian Math. Surveys, **25** (1970), 191–220.

[77] L. V. Keldyš. *Topological imbeddings in Euclidean space*. American Mathematical Society, Providence, R.I., 1968. MR38:696.

[78] Robion C. Kirby and Laurence C. Siebenmann. *Foundational essays on topological manifolds, smoothings, and triangulations*. Princeton University Press, Princeton, NJ, 1977. With notes by John Milnor and Michael Atiyah, Annals of Mathematics Studies, No. 88.

[79] Ulrich Krengel. Entropy of conservative transformations. *Z. Wahrscheinlichkeitstheorie und Verw. Gebiete*, 7:161–181, 1967. MR36:1608.

[80] Peter D. Lax. Approximation of measure preserving transformations. *Comm. Pure Appl. Math.*, 24:133–135, 1971. MR42:7864.

[81] Patrice Le Calvez. Une généralisation du théorème de Conley–Zehnder aux homéomorphismes du tore de dimension deux. *Ergodic Theory Dynam. Systems*, 17(1):71–86, 1997. MR1440768.

[82] Yong Li and Zheng Hua Lin. A constructive proof of the Poincaré–Birkhoff theorem. *Trans. Amer. Math. Soc.*, 347(6):2111–2126, 1995. MR96c:58146.

[83] D. A. Lind and J.-P. Thouvenot. Measure-preserving homeomorphisms of the torus represent all finite entropy ergodic transformations. *Math. Systems Theory*, 11(3):275–282, 1977/78. MR58:28434.

[84] Jie Hua Mai. Besicovitch's topological transformations of plane and Birkhoff's conjecture. *Kexue Tongbao (English Edn)*, 33(7):544–547, 1988. MR90d:58082.

[85] R. Daniel Mauldin, editor. *The Scottish Book*. Birkhäuser, Boston, MA, 1981. Mathematics from the Scottish Café, Including selected papers presented at the Scottish Book Conference held at North Texas State University, Denton, TX, May 1979.

[86] Deane Montgomery. Measure preserving homeomorphisms at fixed points. *Bull. Amer. Math. Soc.*, 51:949–953, 1945. MR7:216b.

[87] Jürgen Moser. On the volume elements on a manifold. *Trans. Amer. Math. Soc.*, 120:286–294, 1965. MR32:409.

[88] J. C. Oxtoby and S. M. Ulam. Measure-preserving homeomorphisms and metrical transitivity. *Ann. of Math. (2)*, 42:874–920, 1941. MR3:211b.

[89] John C. Oxtoby. Note on transitive transformations. *Proc. Nat. Acad. Sci. USA*, 23:443–446, 1937.

[90] John C. Oxtoby. Approximation by measure-preserving homeomorphisms. In *Recent advances in topological dynamics (Proceedings of Conference in Topological Dynamics, Yale Univ., New Haven, Conn., 1972; in honor of Gustav Arnold Hedlund)*, volume 318 of *Lecture Notes in Mathematics*, pages 206–217. Springer-Verlag, Berlin, 1973. MR53:8385.

[91] John C. Oxtoby. *Measure and category. A survey of the analogies between topological and measure spaces*, volume 2 of *Graduate Texts in Mathematics*. Springer-Verlag, New York, second edition, 1980.

[92] John C. Oxtoby and Vidhu S. Prasad. Homeomorphic measures in the Hilbert cube. *Pacific J. Math.*, 77(2):483–497, 1978. MR80h:28006.

[93] Karl Petersen. *Ergodic theory*, volume 2 of *Cambridge Studies in Advanced Mathematics*. Cambridge University Press, Cambridge, 1989. Corrected reprint of the 1983 original; MR92c:28010 and MR87i:28002.

[94] Henri Poincaré. Sur une théorème de géometrie. *Rc. Circ. mat. Palermo.*, 23:375–407, 1912.

[95] Mark Pollicott and Michiko Yuri. *Dynamical systems and ergodic theory*. Cambridge University Press, Cambridge, 1998.

[96] V. S. Prasad. Ergodic measure preserving homeomorphisms of R^n. *Indiana Univ. Math. J.*, 28(6):859–867, 1979. MR80m:28014a.

[97] V. S. Prasad. A mapping theorem for Hilbert cube manifolds. *Proc. Amer. Math. Soc.*, 88(1):165–168, 1983.

[98] A. A. Prikhodcko and V. V. Ryzhikov. The Rokhlin–Halmos–Alpern maximal lemma. *Vestnik Moskov. Univ. Ser. I Mat. Mekh.*, 1996(3):37–41, 92, 1996. MR98k:28026.

[99] Frank Quinn. Ends of maps. III. Dimensions 4 and 5. *J. Differential Geom.*, 17(3):503–521, 1982. MR84j:57012.

[100] V. I. Rokhlin. A "general" measure preserving transformation is not mixing. *Dokl. Akad. Nauk SSSR*, 60:349–351, 1948.

[101] Usha Sachdeva. On category of mixing in infinite measure spaces. *Math. Systems Theory*, 5:319–330, 1971. MR46:5585.

[102] Fritz Schweiger. *Ergodic theory of fibred systems and metric number theory*. Oxford Science Publications. The Clarendon Press Oxford, University Press, New York, 1995.

[103] S. M. Ulam. *Adventures of a mathematician*. Charles Scribner's Sons, New York, 1976. MR58:4954.

[104] John von Neumann. A certain zero-sum two-person game equivalent to the optimal assignment problem. In H. W. Kuhn, editor, *Contributions to the Theory of Games, Vol. II*, volume 28 of *Annals of Mathematics Studies*, pages 5–12. Princeton University Press, Princeton, New Jersey, 1953. MR 14,998i.

[105] John von Neumann. Comparison of cells. In A. H. Taub, editor, *Collected works. Vol. II: Operators, ergodic theory and almost periodic functions in a group*. Pergamon Press, New York, 1961. Unpublished handwritten manuscript with an abstract reviewed by Gilbert Hunt on page 558.

[106] B. Weiss. A note on topological measure theory. *Unpublished*, 1977.

[107] H. E. White Jr. The approximation of one–one measurable transformations by measure preserving homeomorphisms. *Proc. Amer. Math. Soc.*, 44:391–394, 1974. MR51:3401.

[108] Xiao Quan Xu. Besicovitch's planar topological transformations and construction of transitive automorphisms on the closed unit square. *Sichuan Daxue Xuebao*, 27(1):27–28, 1990. MR91h:54060.

[109] Xiao Quan Xu. Explicit transitive automorphisms of the closed unit square. *Proc. Amer. Math. Soc.*, 109(2):571–572, 1990. MR90i:54089.

Index

213